Legal informa

Author and Editc
A946
Email: 3d

The complete imprint of the book can be found on the last pages!

Thank you so much for choosing this book!

Foreword

Thank you very much for choosing this book!

Would you like to deepen your knowledge and skills in CAD design with the software "FreeCAD"? Then this is the book for you! Because on the basis of nine moderately difficult design projects, you can learn new approaches and new features of the program "FreeCAD" in this practice-oriented course and improve your CAD skills in this way.

This course is right for you if you already have beginner knowledge in "FreeCAD" and / or have completed the beginner's course. If this does not apply to you, then please take a look at the beginner's course first. You can find it by searching for "FreeCAD | Step by Step".

I am an engineer and in this course I will teach you advanced design in "FreeCAD" in a simple and easy to understand way.

Here is the link to the free download:

https://www.freecadweb.org

This detailed and practical course is aimed specifically at advanced users and shows in detail and step by step how complex CAD designs succeed. Get your copy now and start improving your skills in FreeCAD today!

Table of Contents

1 Introduction: course scope and software

1.1 What to expect and what you will learn in this course

Hello and welcome to the "FreeCAD" advanced course!

Thank you for choosing this course!

In this course, you will find nine great design projects with easy to medium difficulty. You can reconstruct these projects step by step in the free CAD software "FreeCAD" and in this way improve your CAD skills. As an advanced user, you don't need a big introduction to the program, but certainly want to get started right away. Therefore, after a short note on how to download the program and a few important basic settings, we start immediately with the first design project.

This course is specifically for advanced CAD design only. We will design single parts as well as create assemblies in this course. For the other working areas of "FreeCAD" you will find separate courses for each of them in the course of time. So, the focus of this course is on advanced 2D/3D design with "FreeCAD"!

In this course we will deal with easy projects, such as the construction of a coil spring and a carabiner, as well as with moderately difficult projects, such as the construction of a ball bearing or a remote control. But that was just a small sample, there are many more great projects waiting for you. You will be able to reconstruct each 3D object step by step and one by one in this course, in this way you will consolidate the basic functions in "FreeCAD", but also get to know new functions.

If you have no beginner's knowledge or have never worked with "FreeCAD", you should definitely work through the beginner's course: "FreeCAD | Step by Step" beforehand. This shows you a simple and easy to understand introduction to the program. If you have already completed this course, you are well-prepared for the upcoming design projects!

In brief, this course will teach you in detail:

- Deepen the basic features of "FreeCAD" (from the beginner course) with new projects
- Learn new 2D and 3D features
- Design in a practice-oriented manner using example projects
- Learn new approaches in design
- Create individual parts and assemblies

Design projects:

- *Coil spring,*
- *Carabiner,*
- *Gear,*

- *Flower vase,*
- *Wrench.*
- *Piston, piston pin & connecting rod*
- *Ball bearing,*
- *Watering can,*
- *Remote control.*

It is best to stay in the order given by the course, as the lessons in this course build on each other. Be sure to complete the associated beginner course "FreeCAD | Step by Step", as basics are no longer mentioned in detail in this advanced course. But don't worry, also in this course all details regarding the projects are explained step by step. After a short chapter about the download of the program and the basic settings of the program, we start immediately with the first project!

1.2 Basic settings of "FreeCAD"

"FreeCAD" can be downloaded – as the name already intends – for free. To do this, go to the official website: https://www.freecadweb.org and download the latest version.

Before we can start with the design projects, we will first briefly deal with the settings of the program so that we have an identical starting situation. To do this, click on the button "Edit" and select the option "Preferences...".

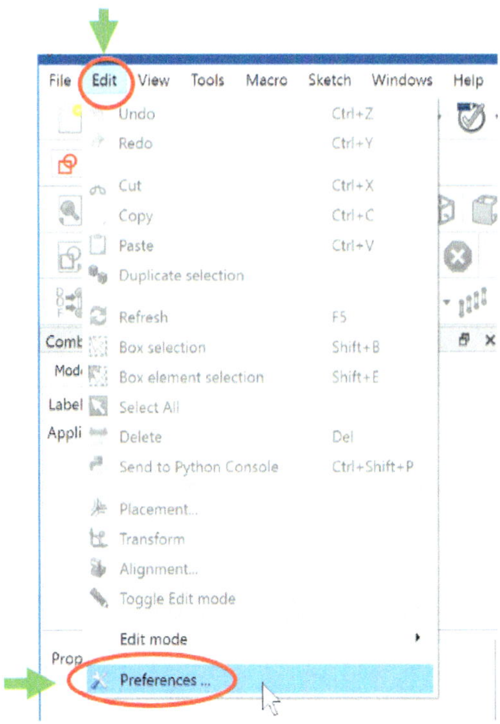

The program "FreeCAD" automatically selects the language of your operating system when you start it for the first time. However, you can also change this setting in the "General" section.

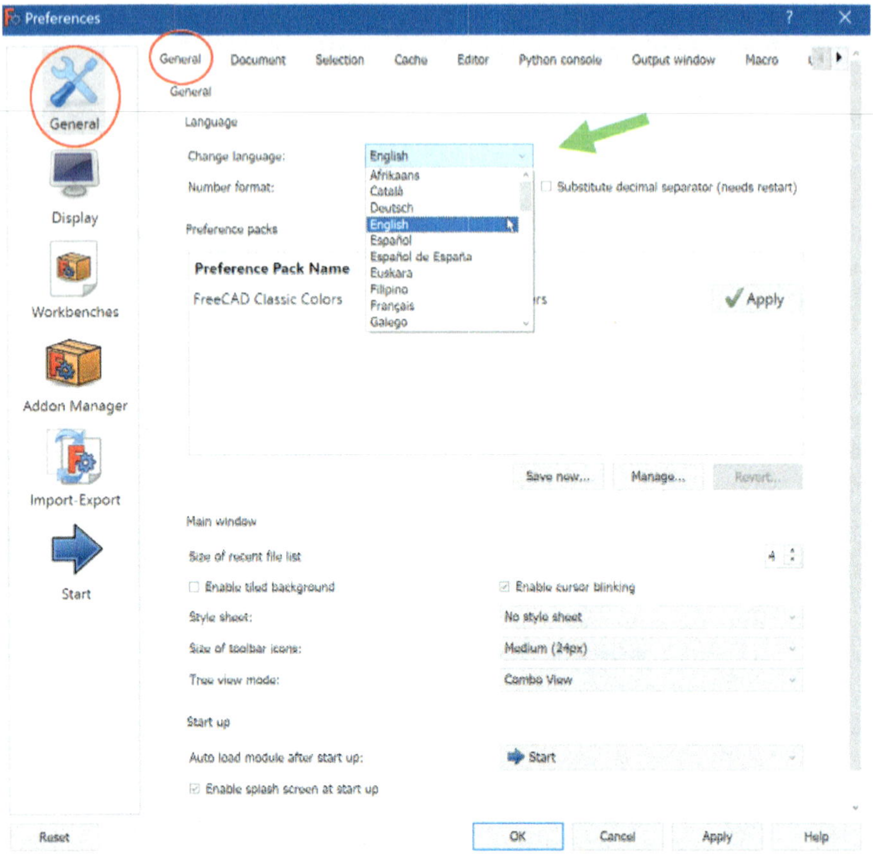

A little further down, in the section "Main window", you can change the color of the display. However, if this is not important to you, you can simply leave the default setting "No style sheet" here. In this section, we can also change the size of the icons for the toolbar commands. It is best to use the "Medium (24px)" setting here if it is not already selected.

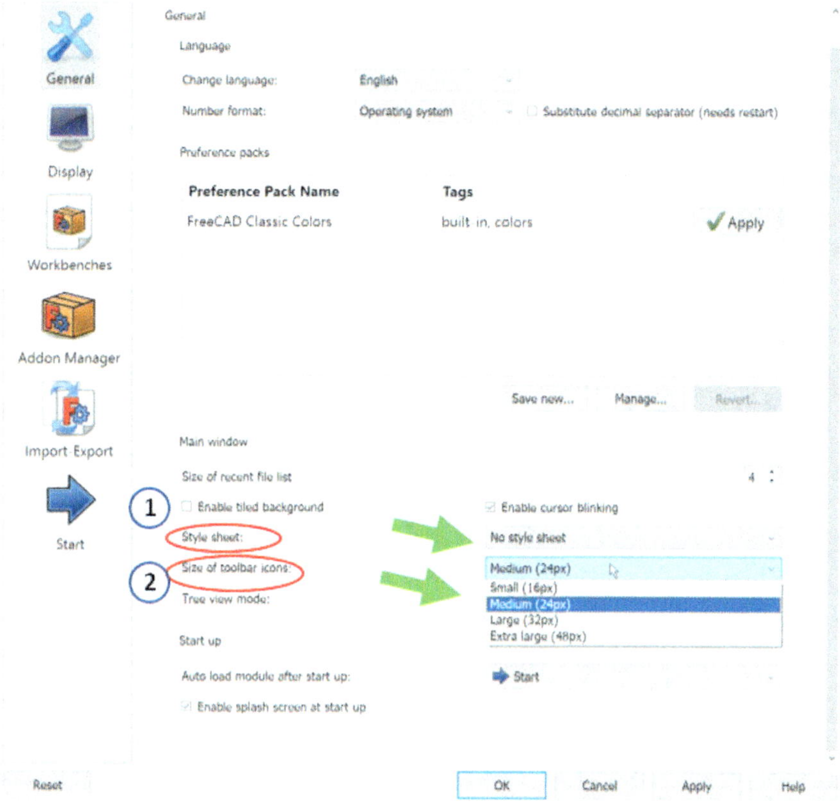

Another important setting from the section "General" we can find in the tab "Units". Here we can set the preferred system of units. We use the standard units "Standard (mm/kg/s/degree)".

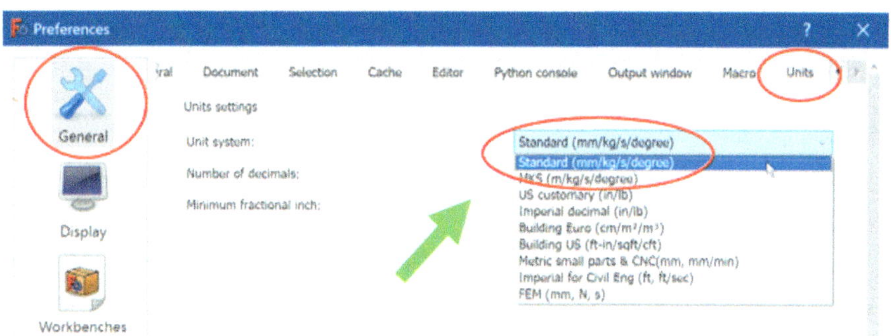

Then we have to check in the area "Display" whether the coordinate system is displayed. For this, the option "Show coordinate system in the corner" must be checked.

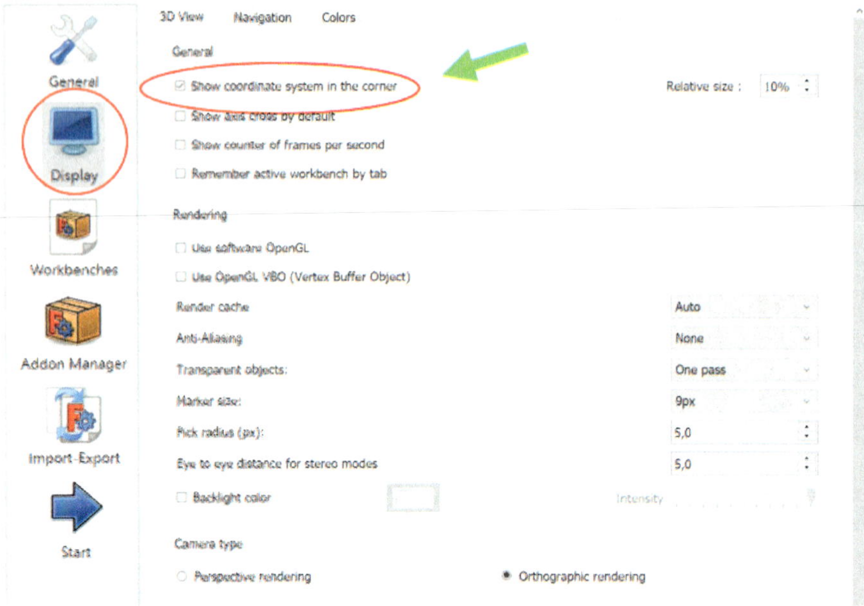

In addition, we can still change the background of the workspace here in the tab "Colors". However, you don't have to do this, it's a matter of taste. For example, we change the background to the color white.

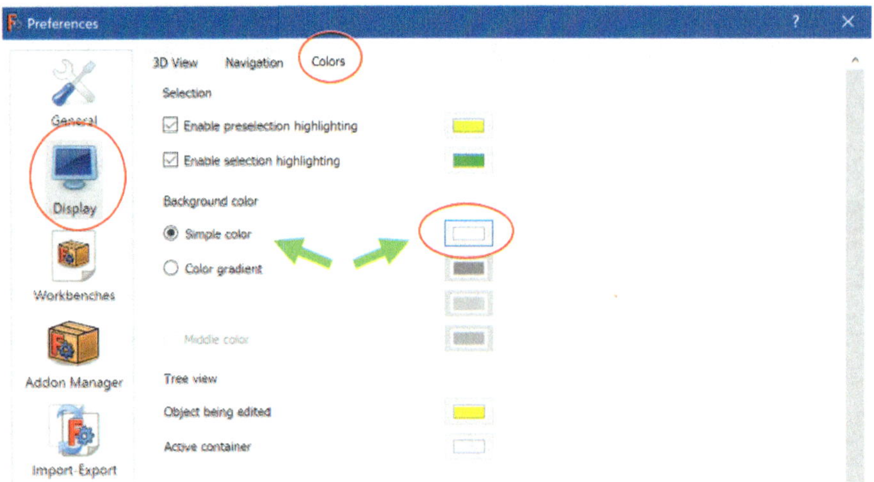

If you have changed settings, click on "Apply" in the lower part of the window and then on the "OK" button to apply the made settings and close the window.

Also, in the workspace "Sketcher" we have to make some important basic settings. As you probably remember, in the workspace "Sketcher" we create the 2D sketches of our 3D objects.

To be able to make the settings, we must first switch to this workspace and then open the preferences ("Preferences ...") again.

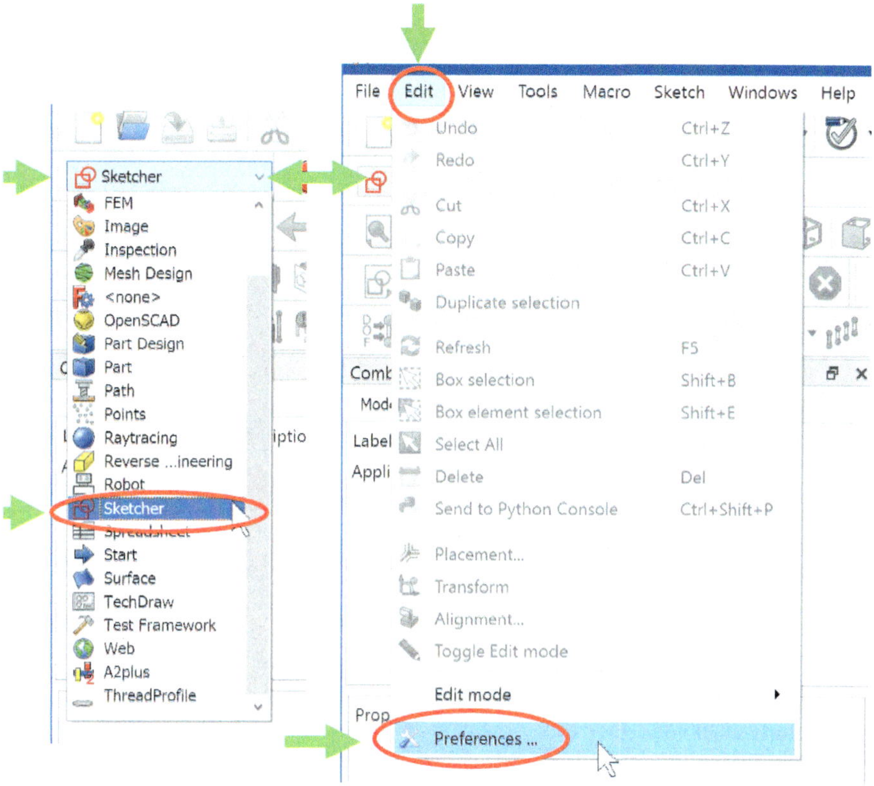

We make a few settings regarding the display. We navigate in the options to the section "Sketcher".

Here we activate – if desired – the drawing grid by setting a check mark at the option "Show grid". The setting "Grid snap" makes it easier for the cursor to select the corner points of the grid. **We will disable this option in this course, as otherwise there could be problems selecting a geometry in the more advanced designs.** The grid size can also be set here. However, these settings are not mandatory, but only an optional help when drawing.

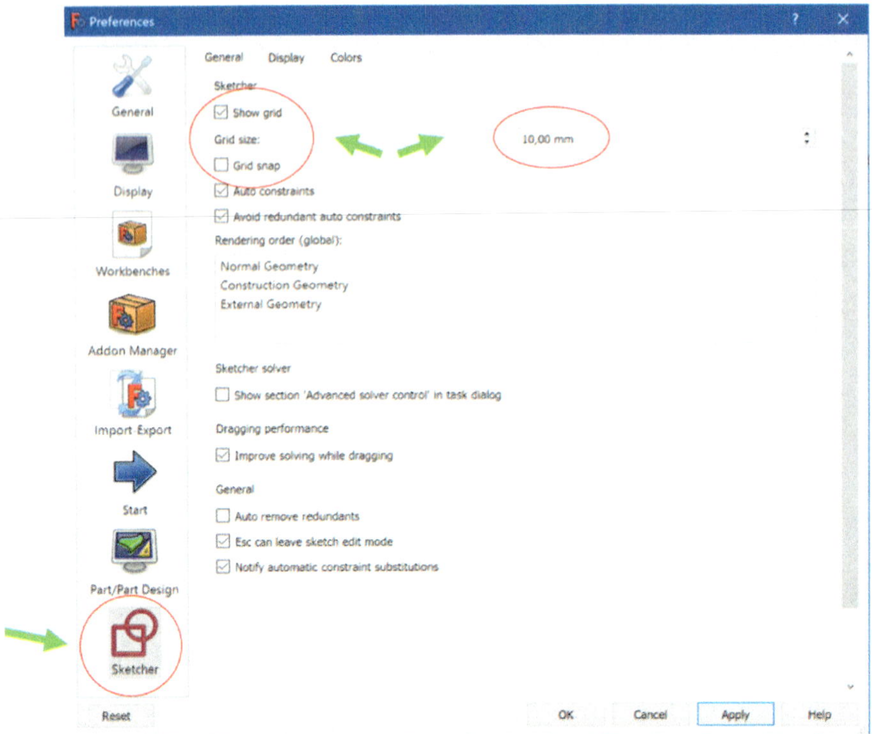

For a better display of the geometric elements, we additionally increase the number of displayed segments per geometry ("Segments per geometry") to the value 500 in the tab "Display". This setting lets e.g., a circle appear rounder. However, this is only noticeable when you zoom in very close to the geometry.

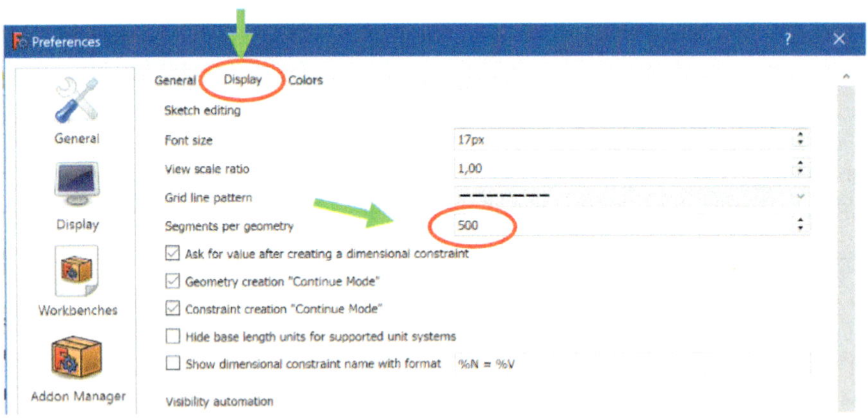

It is also important that in the tab "Colors", we set the fields circled in red to the color black or a similar dark color if we have chosen a white or light background. Otherwise, we would not be able to see the geometric elements later.

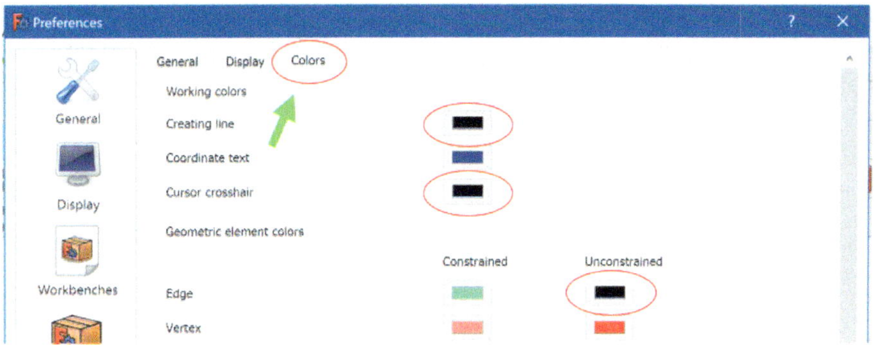

We can also change the display size of the orbit cube ("navigation cube") and the coordinate system in the settings. By default, these are displayed very small. To do this, you need to go to the "Display" section. To make the settings, then first switch to the tab "3D View" and then to the tab "Navigation".

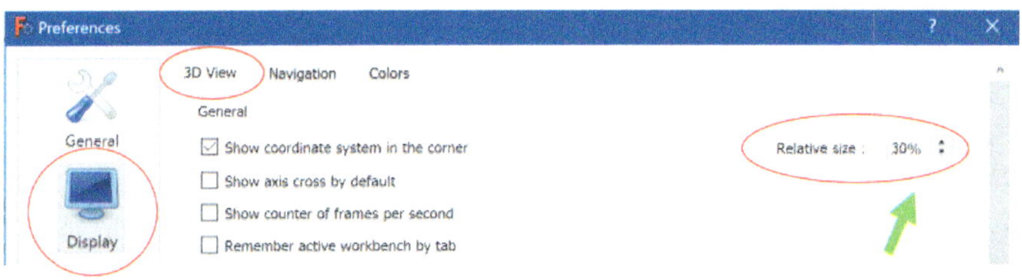

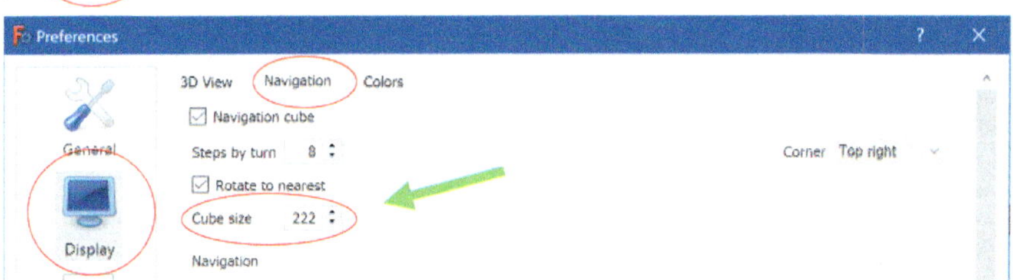

You can move the drawing area and also later the 3D object, depending on the preselected mode. You can select this mode at the bottom right of the drawing area. It is best to select the mode "CAD". The navigation is then as shown. "Pan" means move by the way. "Rotate", "Zoom" and "Select" should be clear.

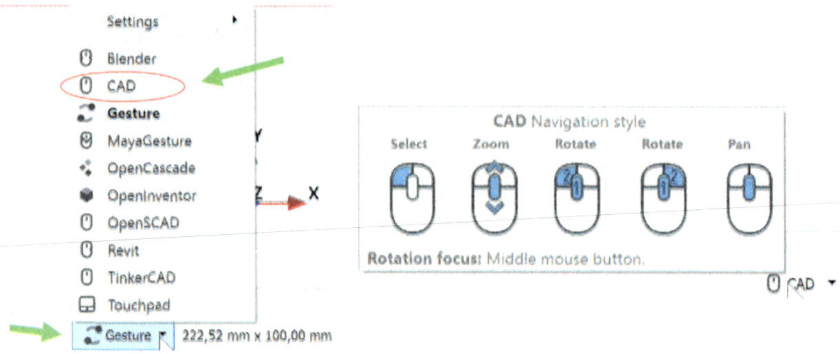

Now all our settings should be identical, and therefore no problems should arise in the further course. We can now finally get started with the design projects!

2 Project No. 1: Helical spring

To warm up, the first project we will create is a coil spring. This should look like this.

At the beginning of each project, we are in the workspace "Part Design". There we create a new document with the command "New" and a body with the command "Create body".

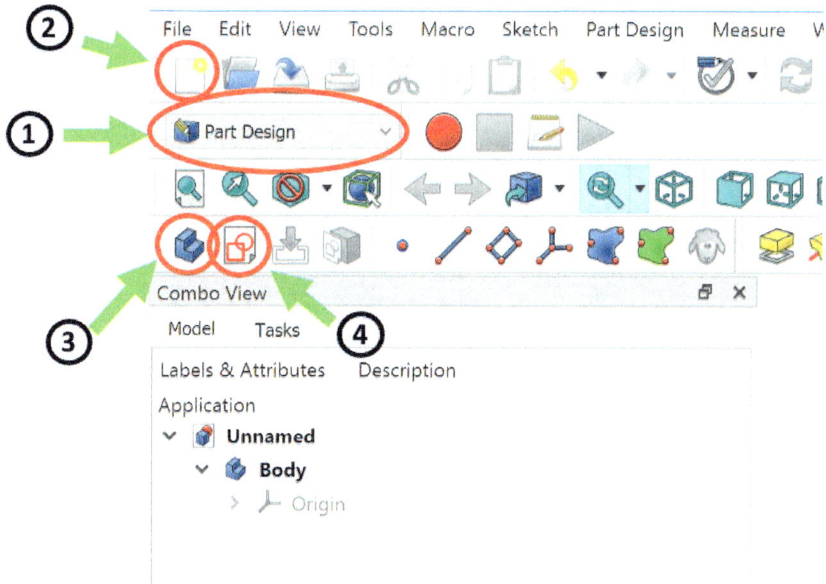

After that, we need to create a 2D sketch. For the helical spring, we create a sketch on the x-z plane using the "Create Sketch" command.

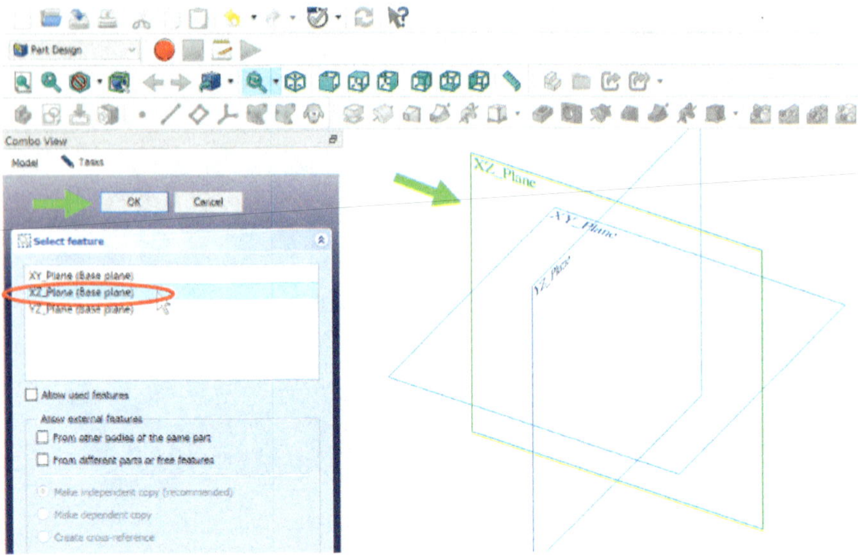

We will create the helical spring using the command "Additive helix". This command extrudes a 2D sketch along a helix. So now we need to figure out what geometry we need to draw in the 2D sketch so that we can then extrude it along a helix and thereby get the helical spring. With a little imagination, we can see that a helical spring consists of a piece of wire that has been wound in a spiral. This means that we need a circle as the initial geometry for the piece of wire.

This circle should sit on the x-axis and have a diameter of 2 mm. This dimension will later determine the material thickness of the helical spring. In addition, the circle is to sit 10 mm away from the coordinate origin. This dimension later determines the diameter (20 mm) of the helical spring.

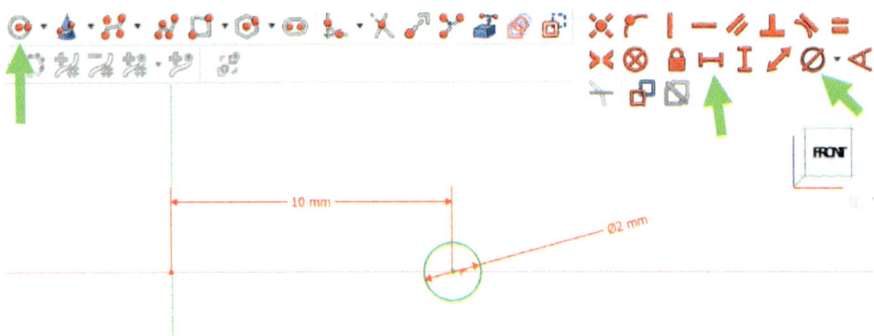

After that, we can close the sketch. After making sure that the sketch is selected in the structure tree, we click on the command "Additive helix".

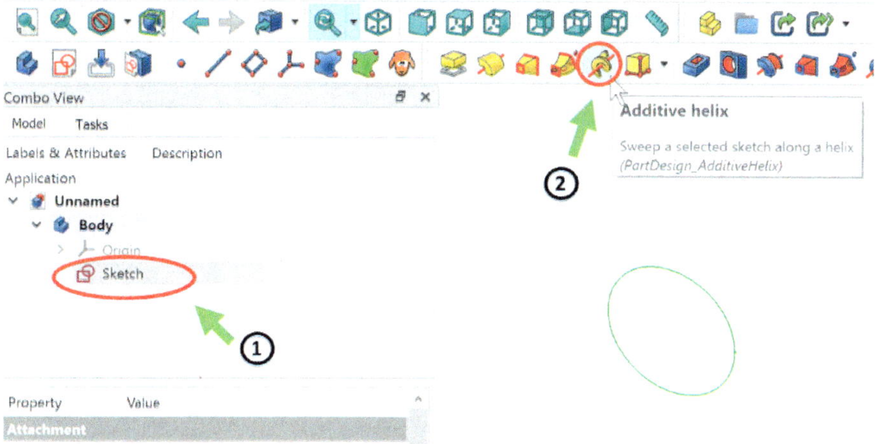

The program then extrudes the sketch along a helix and thus creates the helical spring. The option "Vertical sketch axis" is selected as axis in the settings. Equivalent here would be to select the option "Base z axis" since the z axis is the vertical axis in this case. If we select one of the other axes, the orientation of the coil spring will be changed.

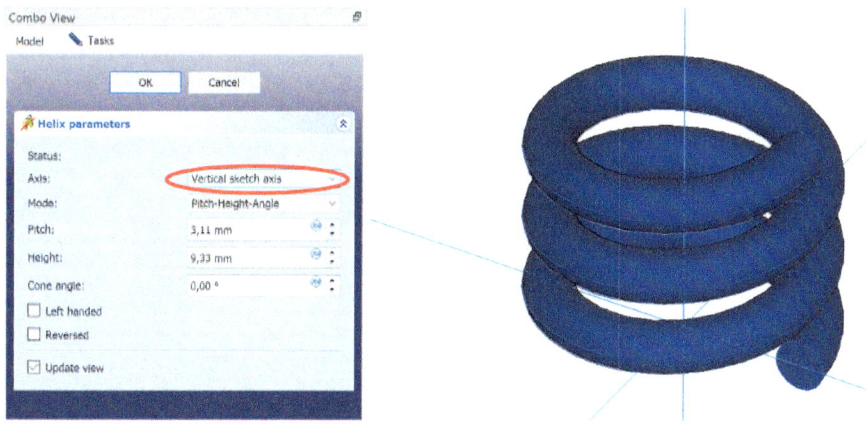

Using three parameters, we can modify the coil spring in a desired way. By default, the option "Pitch-Height-Angle" is selected. In this case, we need the desired height, the desired pitch and, if necessary, an angle if we want to get a conical coil spring. For example, we enter 3.11 mm in the parameter "Pitch", 50 mm in the parameter "Height" and 0° in the parameter "Cone angle". By clicking on "OK", we create the 3D model.

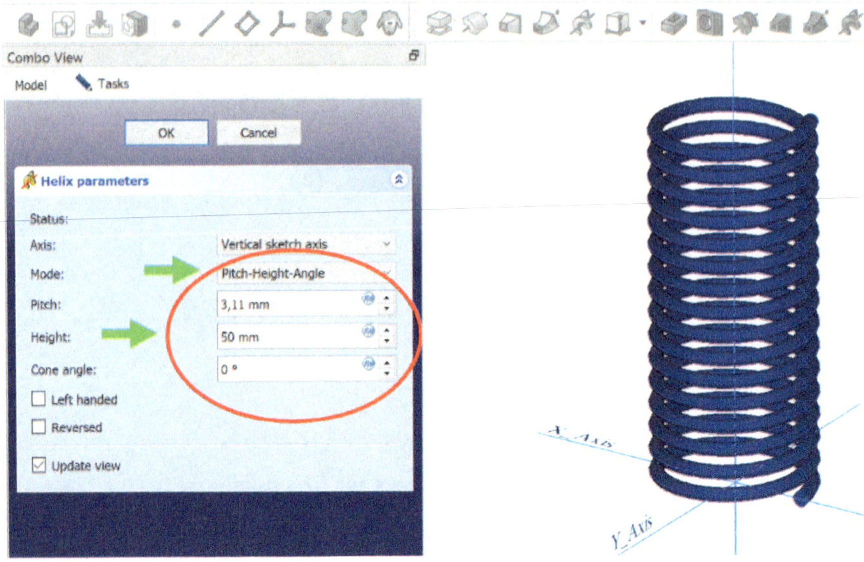

This project was relatively easy and, as mentioned earlier, meant to be a warm-up. Don't worry, the difficulty level increases with each project. There are still many great and more complex projects waiting for you!

3 Project No. 2: Carabiner

Next, we want to design a carabiner that looks like this.

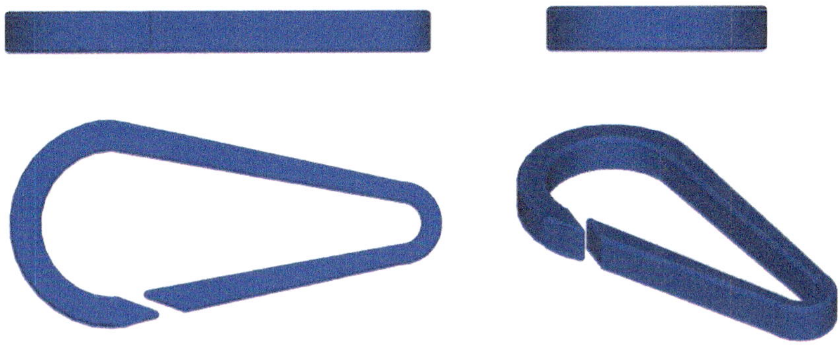

For the carabiner, we create a new document in the workspace "Part Design" as usual. First, we think about the best way to build the 3D model of the carabiner. If we look at the carabiner a little more closely, we notice that we can place a circular shape in each of the left and right areas. Furthermore we can see, that the struts of the carabiner represent tangential connections between these two circles.

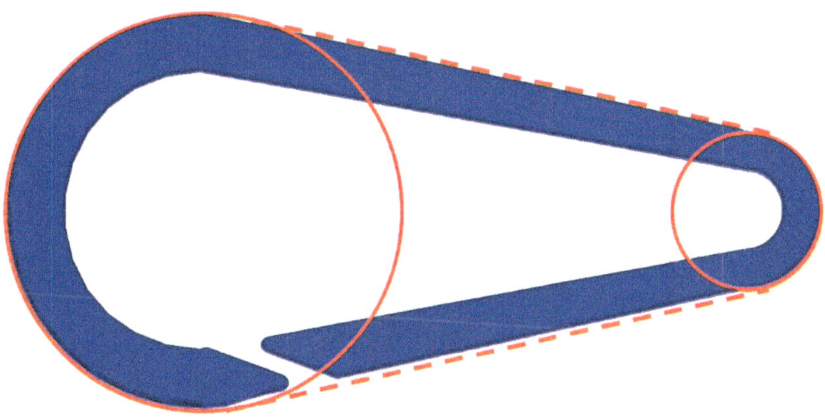

We will construct the carabiner with the help of these geometries. To do this, we first need a 2D sketch, which we will create on the x-y plane.

On this plane, we first draw the two circles. The centers of the two circles should lie on the horizontal red line (x-axis). For example, we choose a diameter of 50 mm for the first circle. Then create another circle with a diameter of 20 mm. The first circle should be on the left side of the green vertical line (y-axis), the second circle should be on the right side of this line. We dimensioned the distance between the two circles as 70 mm. To completely define the previous sketch, we now need a reference to the origin along the x-axis. We define the position of our sketch in x-direction, for example, by dimensioning 35 mm from the center of a circle to the origin.

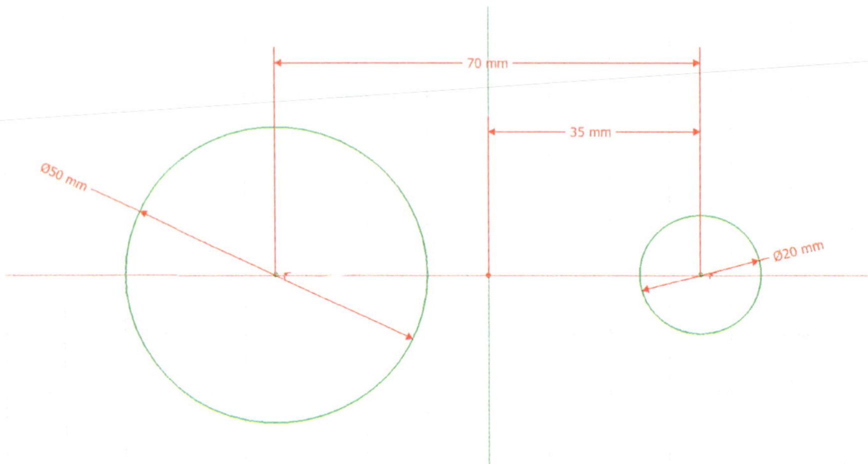

Then we draw two lines whose endpoints should lie on the two circles respectively. We need one line above the x-axis and one line below the x-axis.

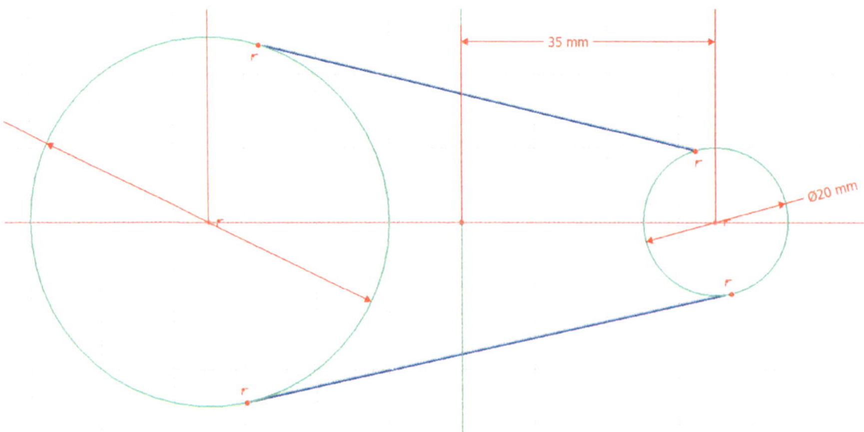

In the next step, we create vertical constraints between the line endpoints and circle centers so that the line endpoints are vertically above the circle centers. To do this, we first select two points at a time and then use the command "Constrain vertically".

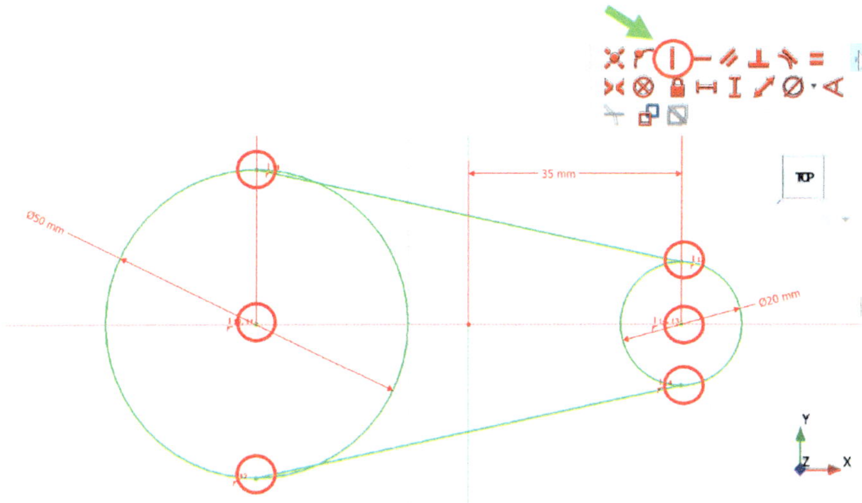

For the further design we need only the outer contour, therefore in the following we use the command "Trim Edge". With this command, we can easily remove all superfluous line segments (marked with arrows). The sketch remains completely defined because the program automatically links the resulting vertices.

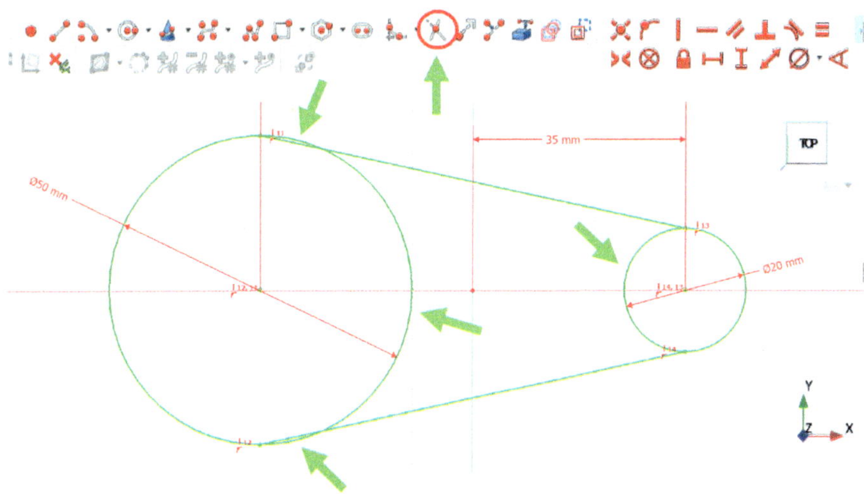

After applying the command, we get the outer contour of the carabiner.

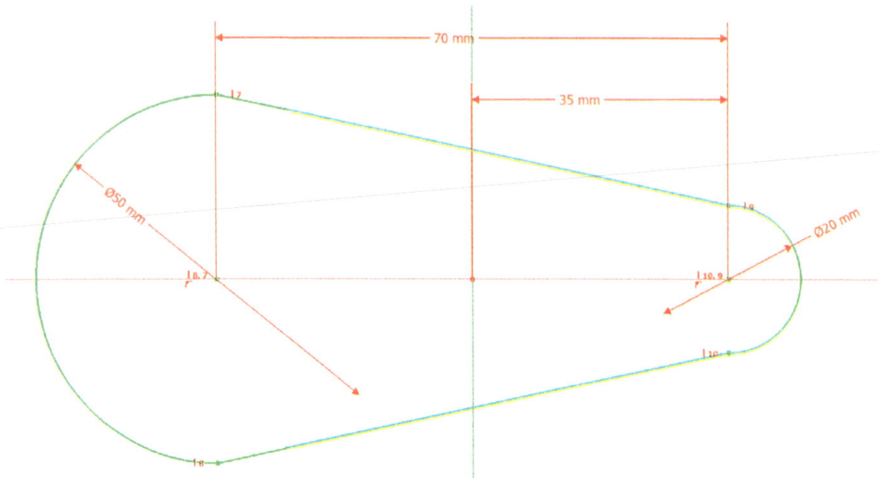

If we were to extrude this surface already now, we would have to make an additional cutout in the center area to get the final base body of the carabiner. However, we can save this step by drawing the cross-section of the carabiner in one step. To do this, add two additional circles in the inner area of the carabiner and dimension them as 35 mm and 10 mm.

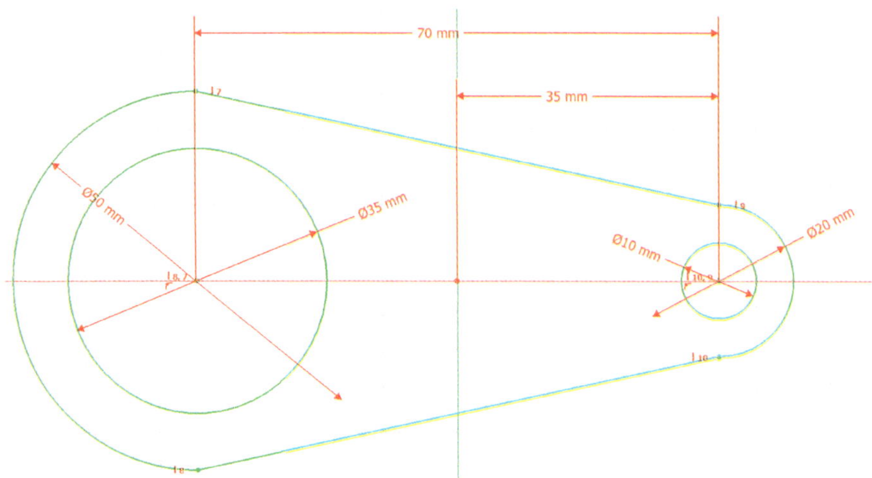

Analogous to the previous steps, we then again create two lines whose endpoints lie on the two circles. In addition, we also create four constraints here with the command "Constrain vertical", so that the end points of the lines coincide in the vertical direction with the two circle centers.

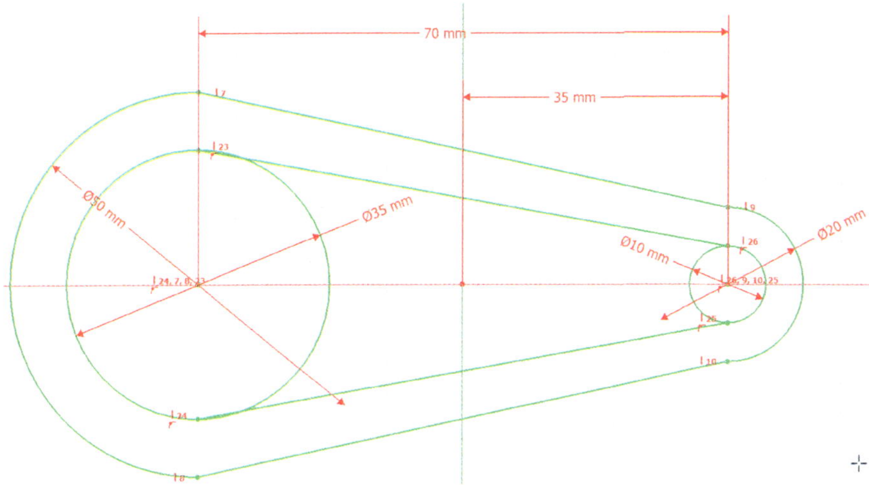

Finally, we remove all superfluous geometry using the "Trim edge" command.

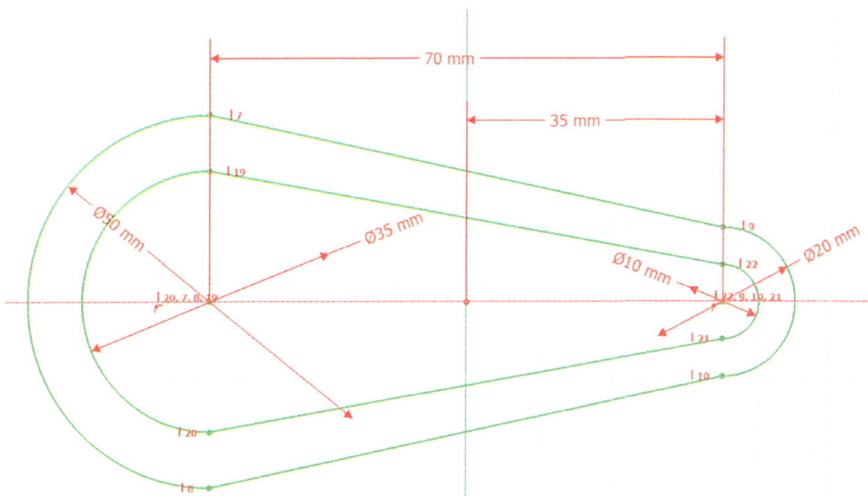

As you can now see, this gives us the almost finished cross-section geometry of the carbine. There is still one opening missing. For this opening, we draw two parallel lines in the lower-left area of the carabiner. The line end points should lie on the contour of the carabiner. We create the parallelism of the lines with the condition "Constrain parallel".

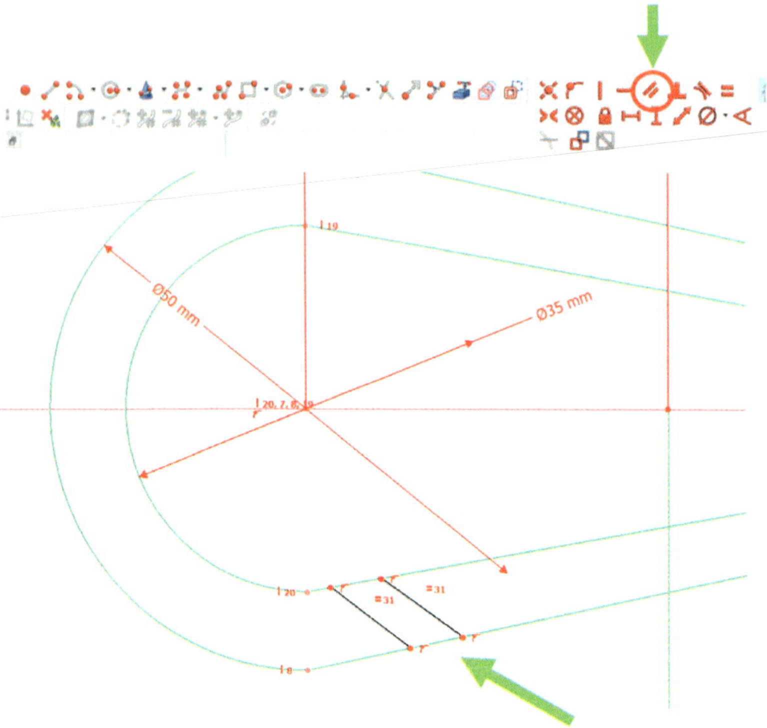

For the further definition of the sketch, we add an angle dimension which we place between the bottom line and the x-axis with the command "Constrain angle". We do this by first clicking on the line, then on the x-axis, and then on the command. We need an angle of 20 degrees.

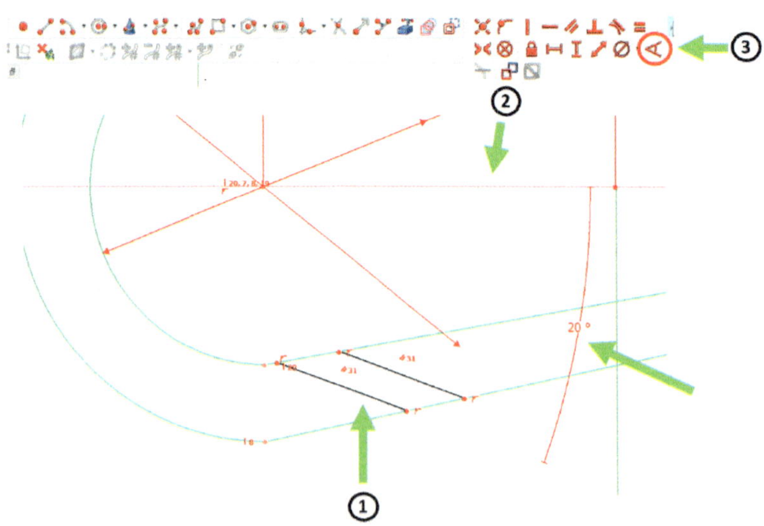

To finally define the sketch completely, we add two horizontal dimensions of 4 mm (between the two parallel lines) and 1 mm (between the end point and the center of the circle).

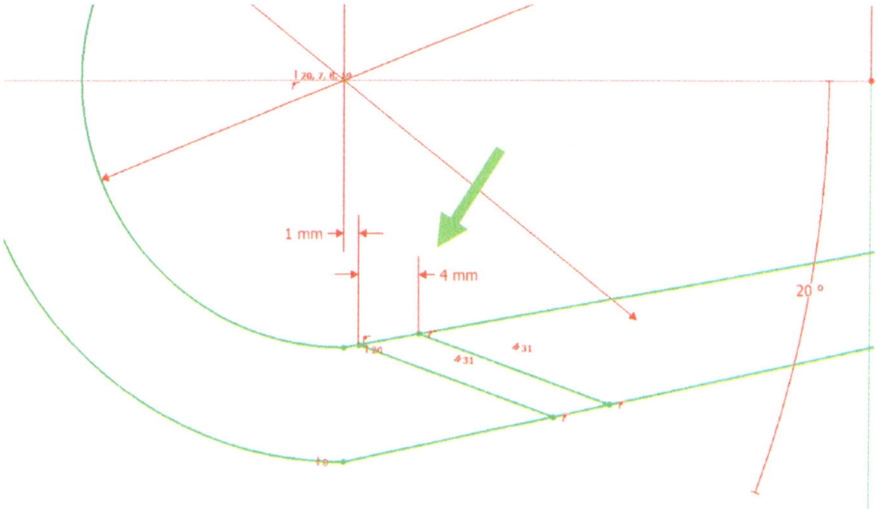

Since we want to construct an opening, we also need to remove two superfluous line segments. We do this again with the command "Trim edge".

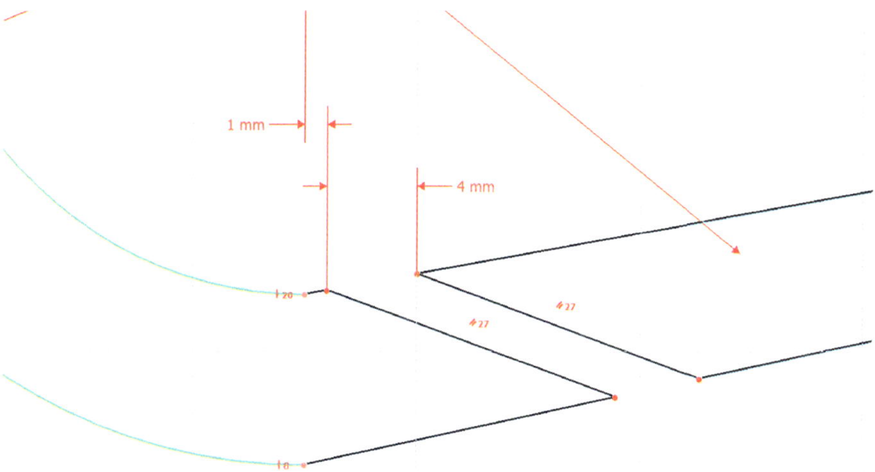

Since the sketch is now unfortunately no longer completely defined by this action (some lines became black), we must add a few dimensions so that the sketch turns completely green again. We do this with the help of the command "Constrain Distance" and a click on the respective lines (see arrows). We leave the dimensions at the displayed values.

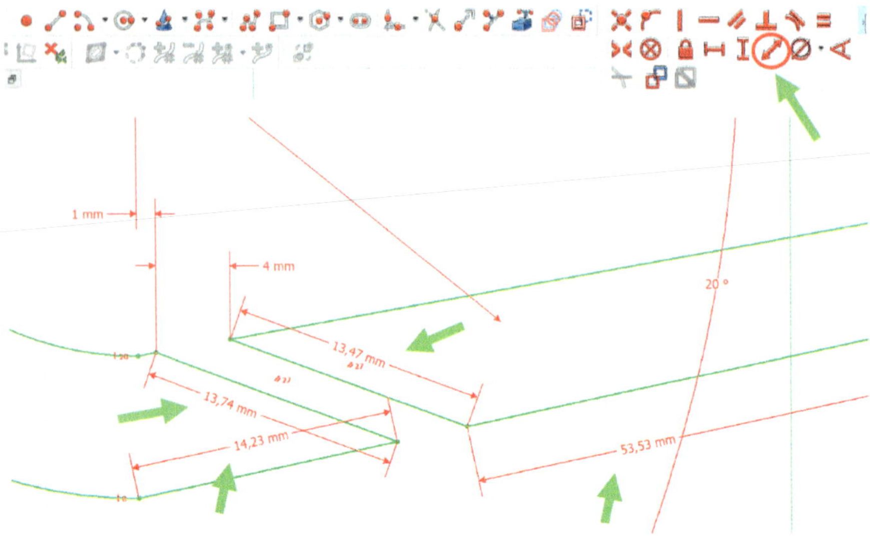

To transform the 2D surface into a 3D body, we switch with the button "Close" from the workspace "Sketcher" back to the workspace "Part Design" and use the function "Pad" there. We enter a value of 10 mm.

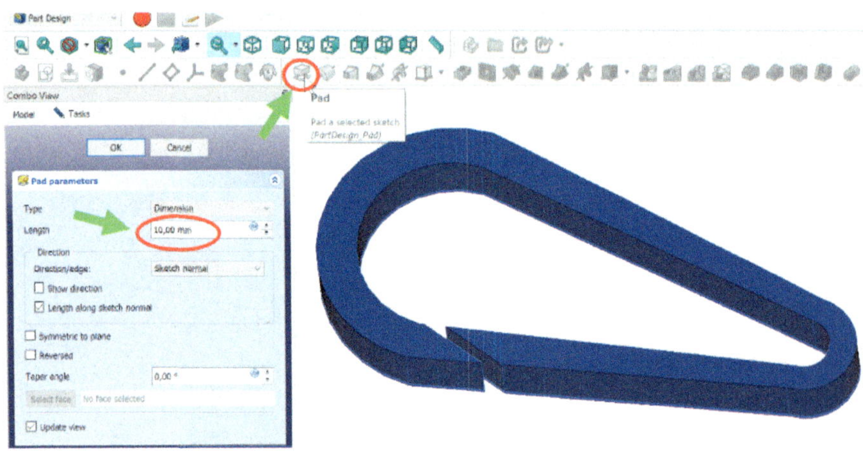

You can extrude either in one direction only, symmetrically, or independently in two directions. You select this at the setting "Type" in the combination view. If you want to have a tapered shape, you could specify an angle at the "Taper Angle" setting. However, we do not need that here. With a click on "OK" we create the 3D body.

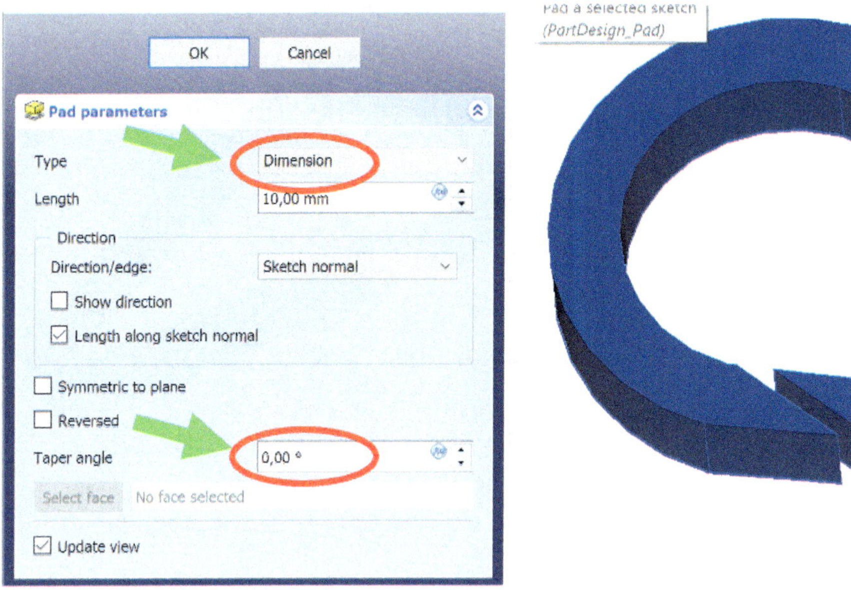

Finally, we round a few edges using the "Fillet" command. For example, we use a radius of 20 mm for the back top edge.

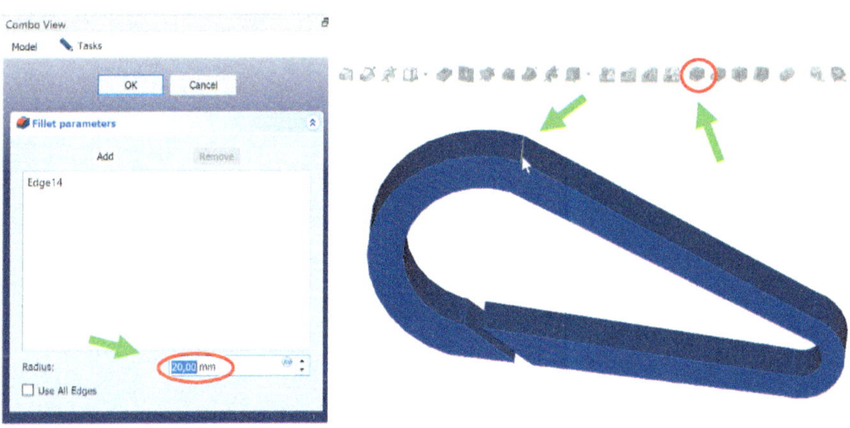

We also round the remaining edges by 1 mm each. To do this, we simply select the top surface of the carabiner, click on the command "Fillet", and then activate the option "Use all Edges" in the settings to round all edges.

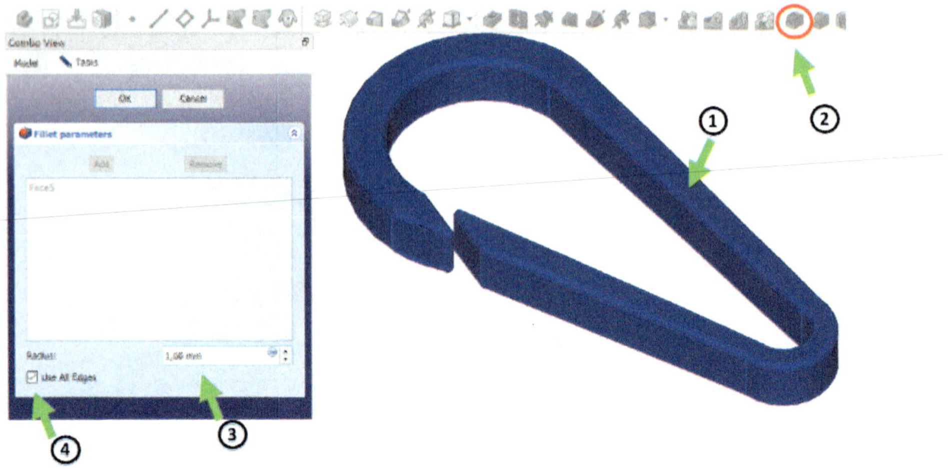

Flawless! Please save the file before we move on to the next design project – a gear.

4 Project No. 3: Gear

The next project we would like to design is a gear, which could be a part of a more complex machine, for example.

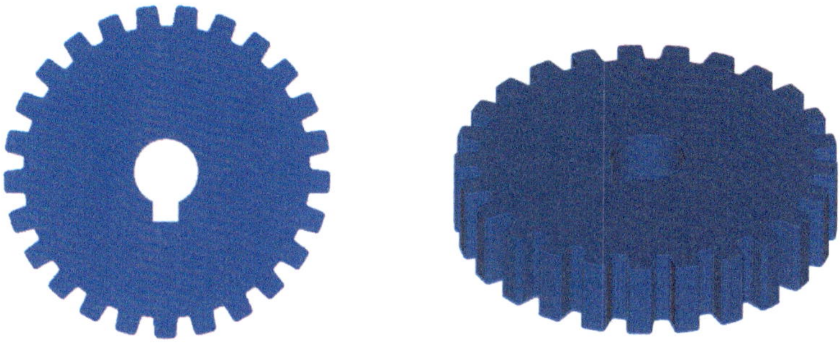

We first create the basic body of the gear, including the cutout in the center, which is used to hold a shaft. We create the cutouts for the teeth of the gear later because, as we will see, this is the most efficient way to work. For the main body, we sketch in a new "Part Design" document on the x-y plane. We first need a circle with 50 mm diameter, whose center should be on the coordinate origin.

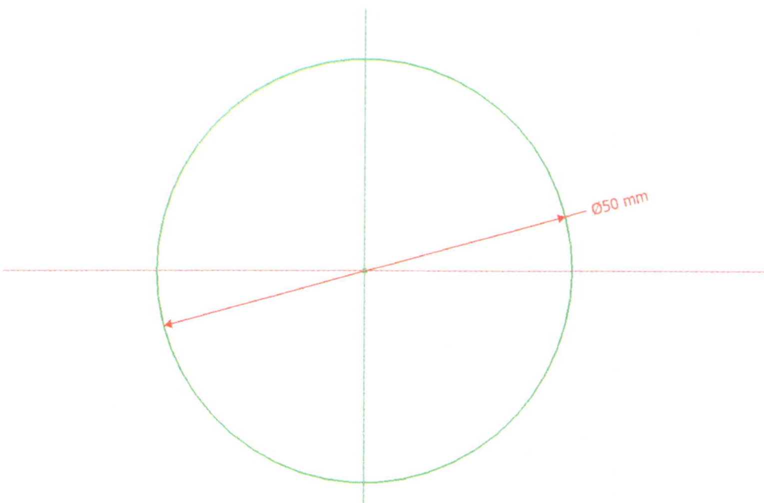

This circle forms the outer contour of the gear.

For the second part of the gear wheel (bore or cutout), with which the gear wheel can later be driven with the aid of a shaft including a drive nose, we first need a circle with a diameter of 10 mm, the center of which should also lie in the coordinate origin.

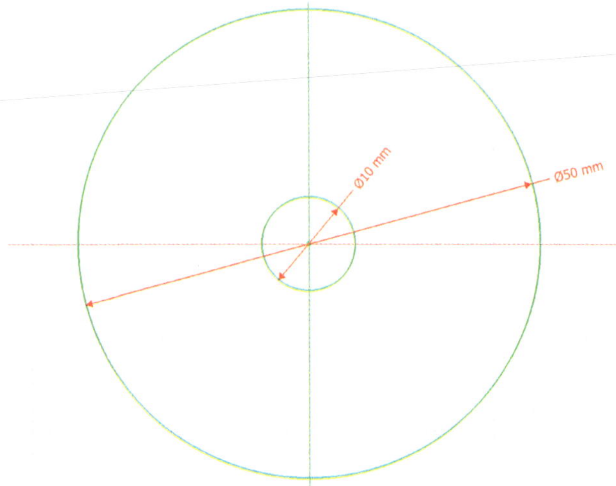

We then create the rectangular cutout for the shaft's drive nose using a 3 mm vertical line, whose starting point should be on the circle, followed by a 4 mm horizontal line and another vertical line that complete the rectangular profile. We also add a 2 mm dimension from one of the two side lines to the origin.

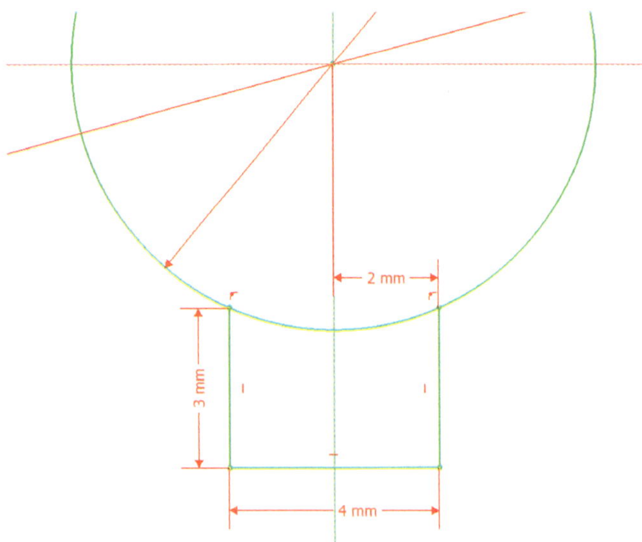

Then we remove the superfluous circle segment with the command "Trim edge", so that a continuous surface is created.

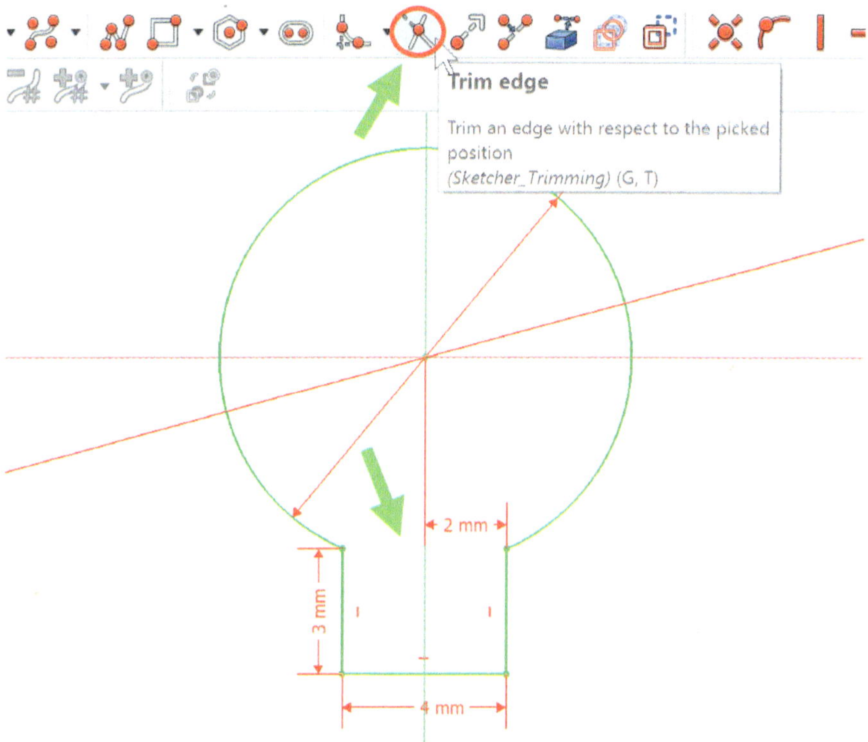

Now the complete base body of the gear is ready and can be extruded using the function "Pad". Before that, of course, we have to close the sketch with the button "Close".

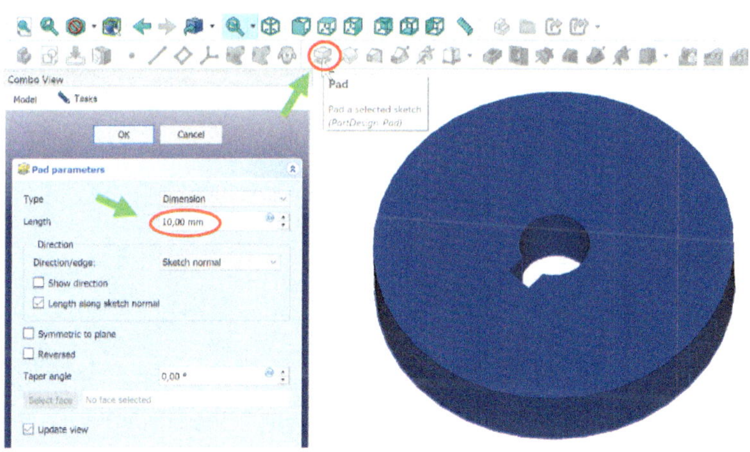

We will now cut out the teeth of the gear from this basic body. To do this, we will sketch the first tooth in a new sketch on the upper surface of the 3D body we have just created. We will then easily duplicate this tooth later using the "Polar Pattern" command so that we do not have to draw all the teeth individually.

After we have created a sketch, we hide the basic body of the gear so that we have a better view while drawing. We do this by switching to the tab "Model" in the combination view, selecting the body ("Body") and pressing the space bar. Now the body has been hidden.

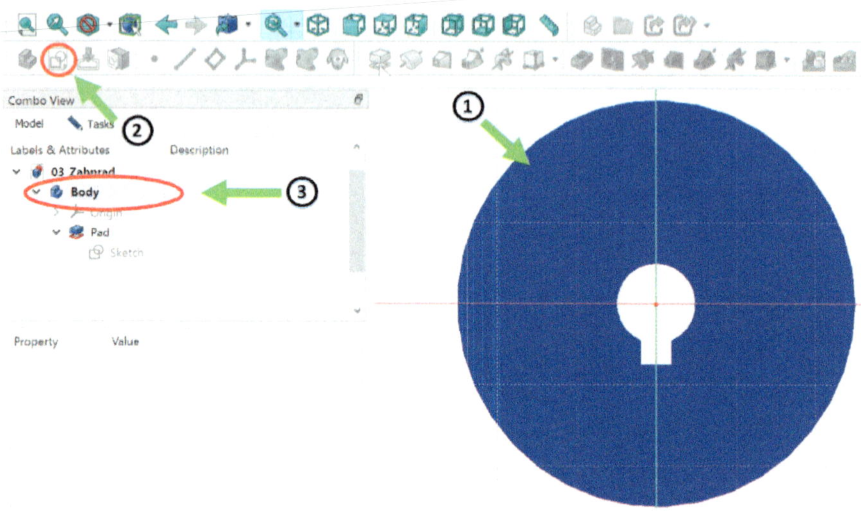

For the contour of the gear, we again draw a circle with a diameter of 50 mm in this sketch.

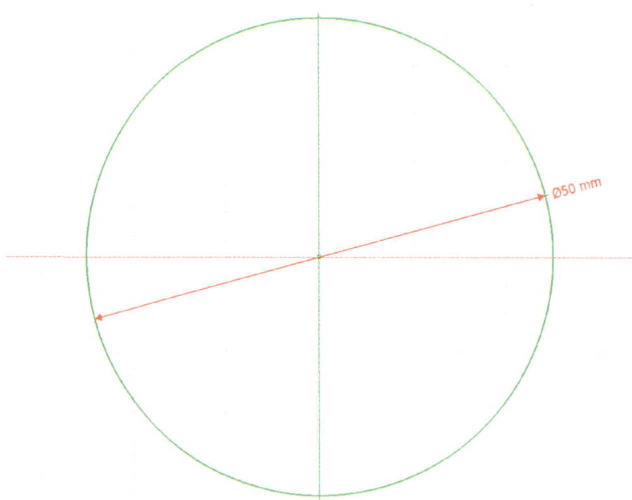

We then first draw only one half of the tooth, which we then simply mirror on the y-axis. We need a 1 mm horizontal line for this, which sits on the y-axis (green, vertical line).

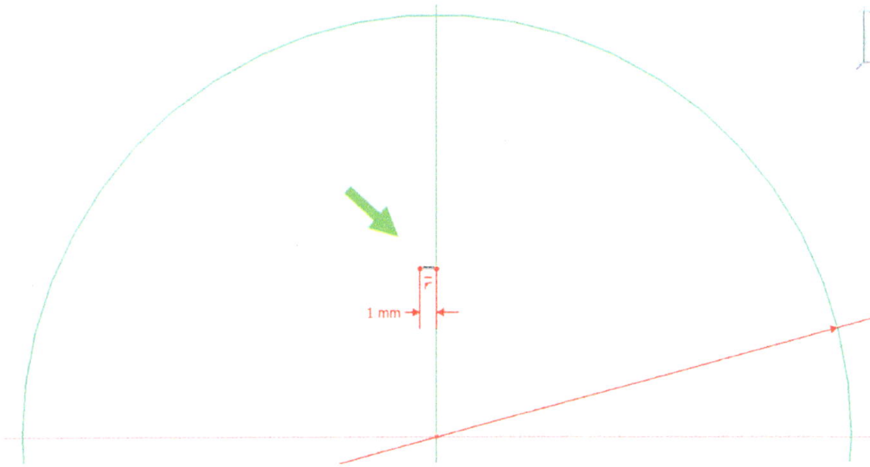

Then follows a second line, which we stretch to the upper left.

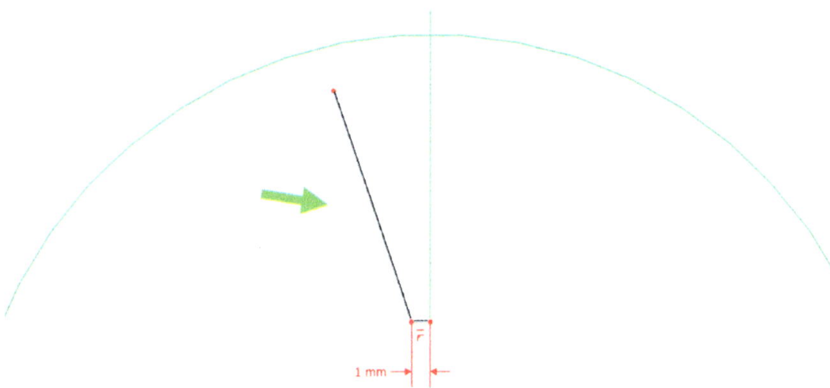

In the upper area we then add a 3-point arc ("End points and rim point"), whose start and end points should sit on the circle on the one hand, and on the end point of the previously created line on the other. It is best to start the arc on the circle.

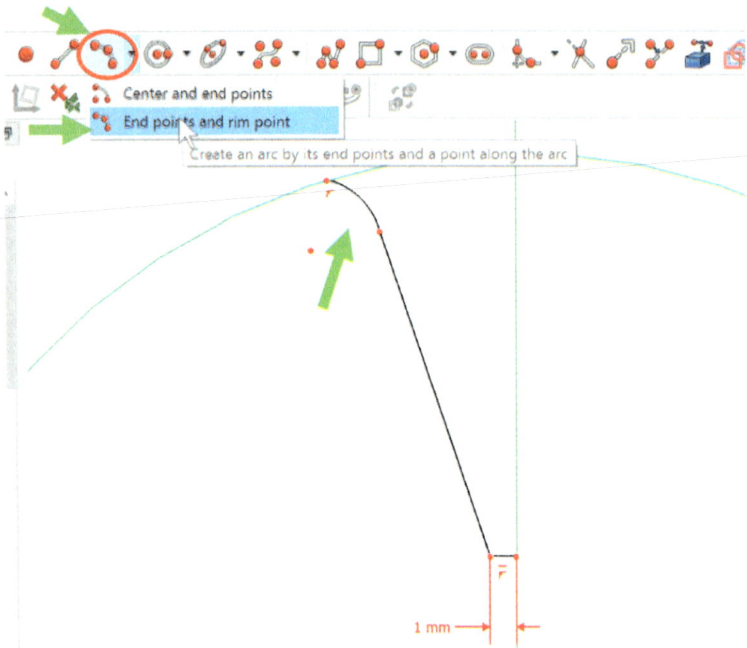

Please ensure that the correct constraints are created. We need a tangential constraint between the 3-point arc and the circle and between the 3-point arc and the line. Add these constraints using the command "Constrain tangent" if they are missing or if there are others instead. You can see this by the small red symbols. If a message appears saying that you are replacing constraints, you can simply ignore it and click away.

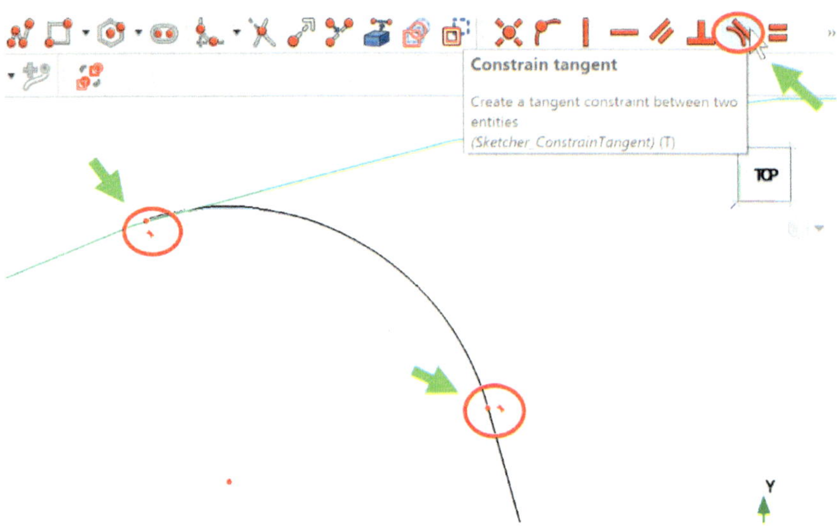

We then dimensioned the distance between the corner point of the tangent arc and the starting point of the first line as 3 mm (vertical direction; "Constrain vertically").

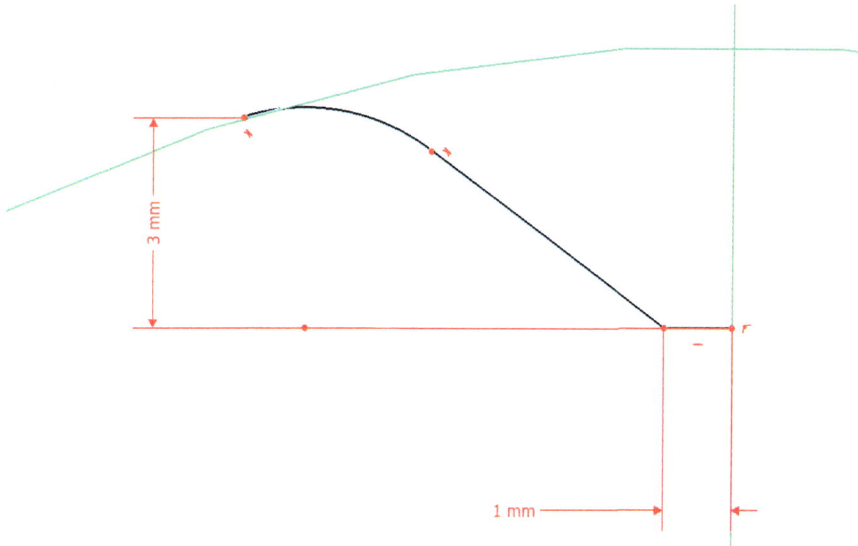

Then we dimensioned the distance between the starting point of the first drawn line and the corner point of the 3-point arc with 2 mm (horizontal direction; "Constrain horizontally").

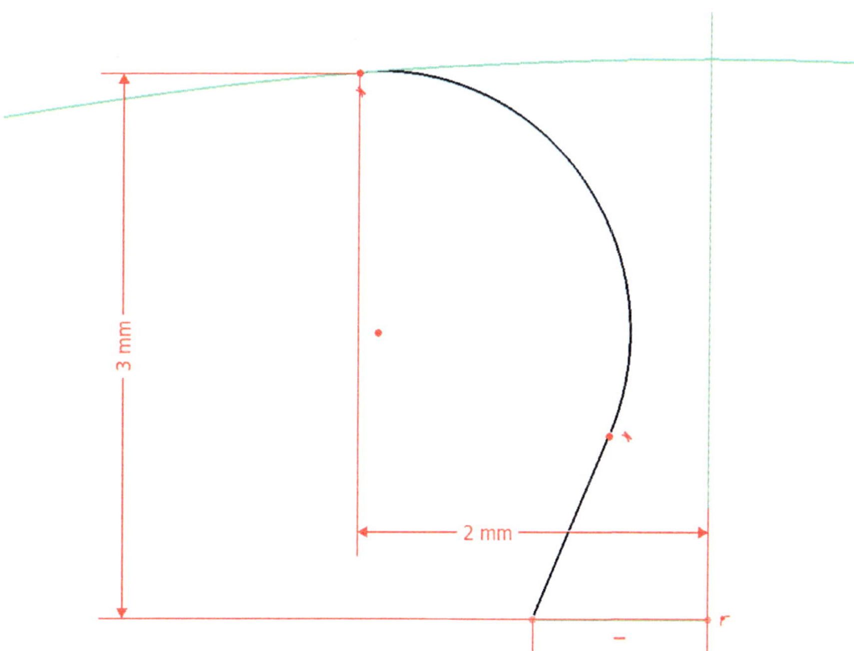

Finally, we define the radius of the 3-point arc as 0.5 mm using the command "Constrain radius".

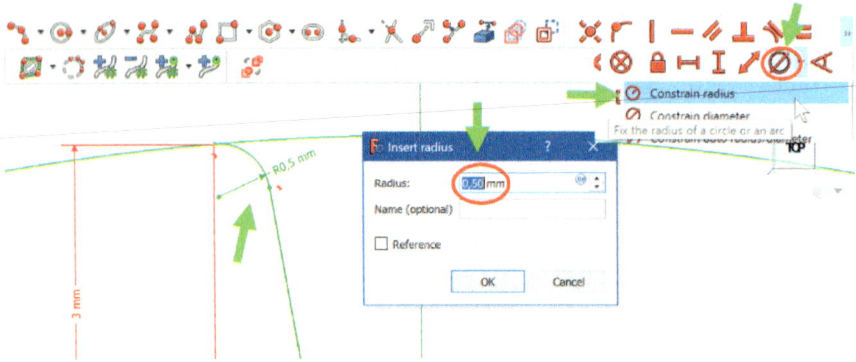

Now the profile can be mirrored. But before we do that, we round the lower-left corner of the profile by using the function "Sketch fillet" and entering a radius of 0.5 mm.

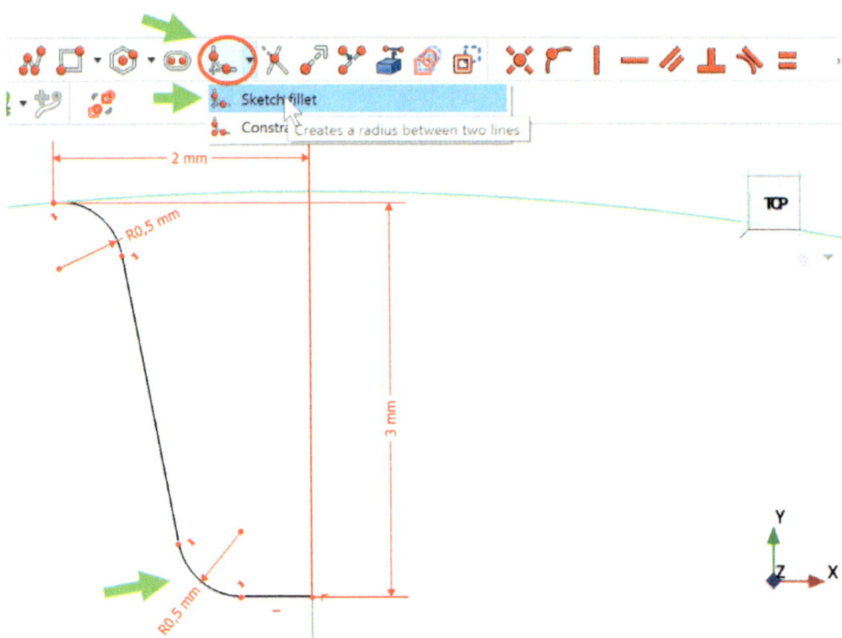

For a complete definition of the sketch, we can dimension the lowest horizontal line with the command "Constrain distance". The value for the dimension in this case must be 0.56 mm.

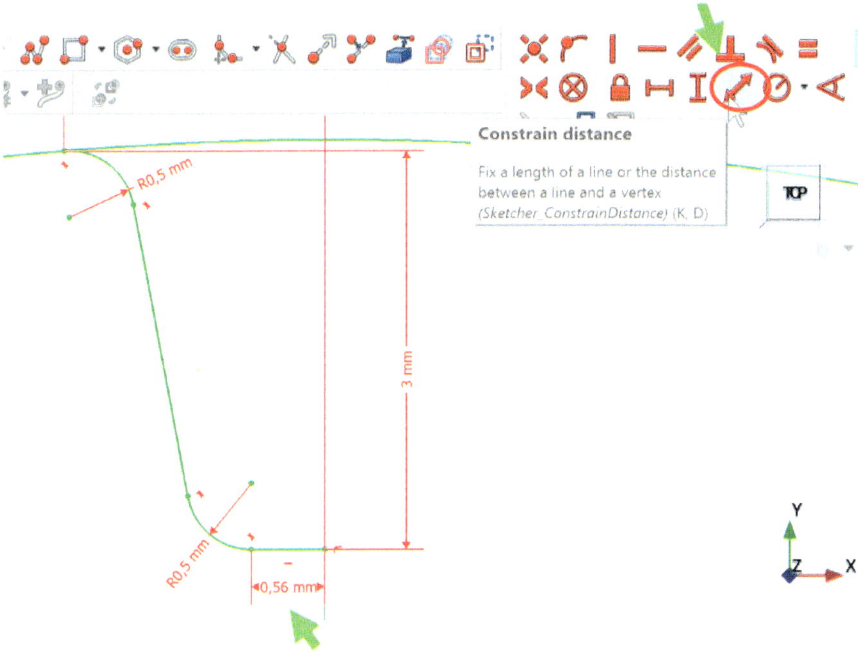

Now we want to mirror this profile to the other side of the y-axis to get the geometry for the first tooth of the gear. To do this, we click successively on the elements that we want to mirror and lastly on the y-axis that will serve as the reference for the mirroring. Then we select the command "Symmetry" from the toolbar.

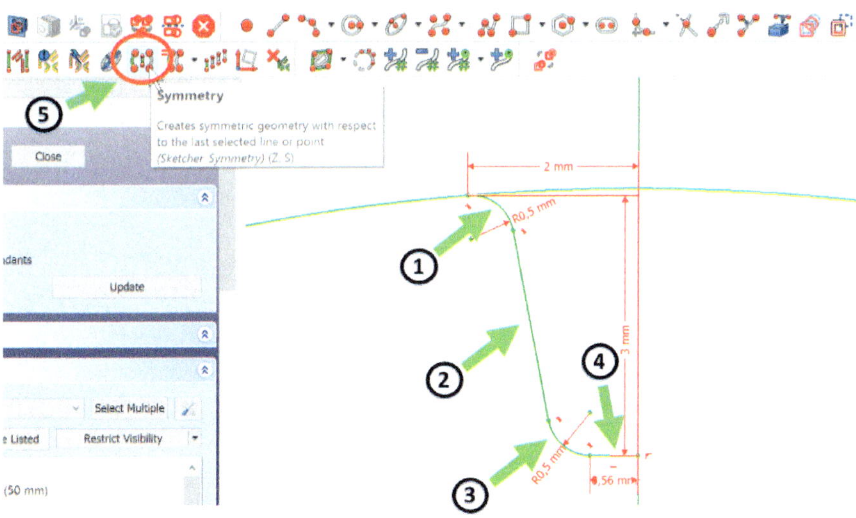

This gives us the second half of the profile.

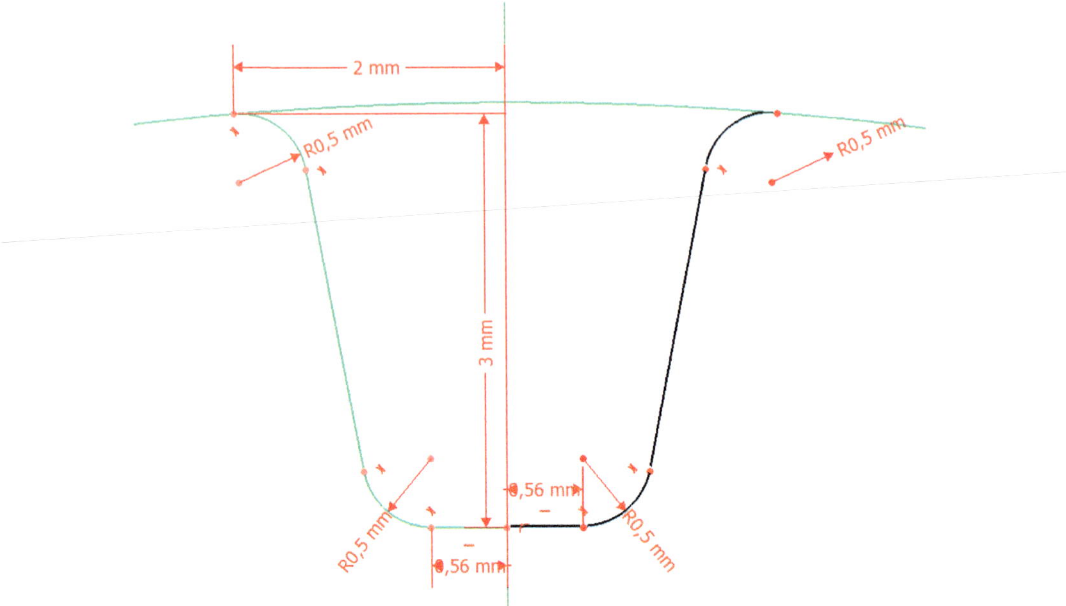

If we drag the geometry with the right mouse button pressed, we notice that we can still move it freely. We can also see this from the black color of the geometry. To define it completely, we have to add relevant constraints and dimensions. Try this out on your own. If you get stuck, you can use the instructions for the first half (a few pages back) as a guide, since the constraints and dimensions are identical for both halves.

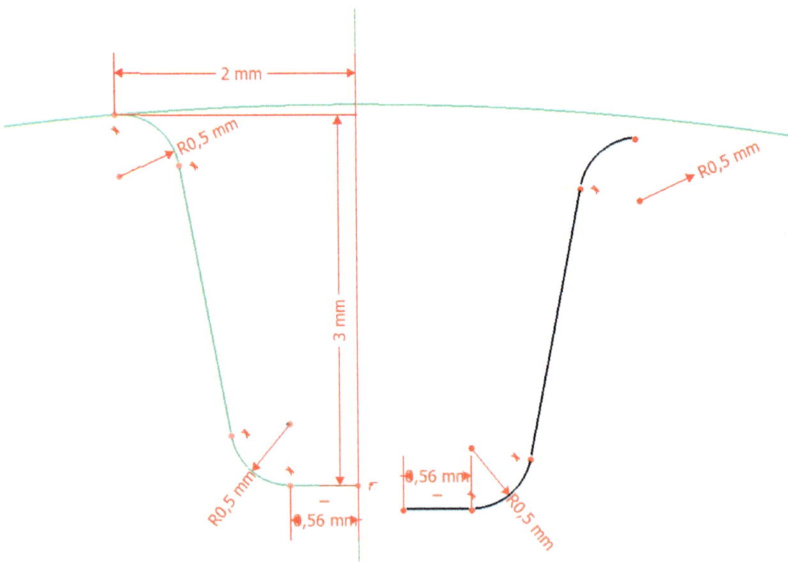

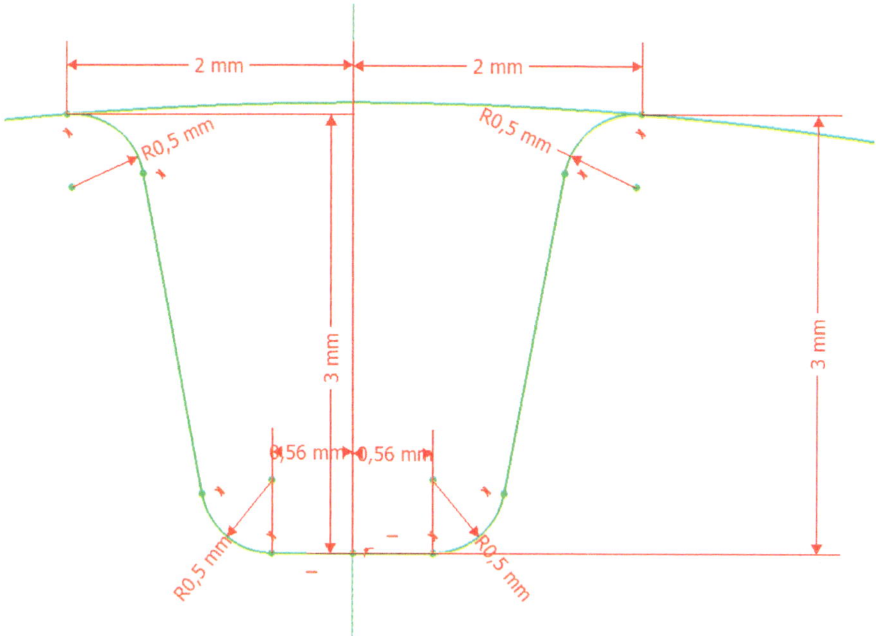

The profile for the first tooth is ready. Before we close the sketch, however, we remove the outer contour of the 50 mm circle by using the function "Trim edge".

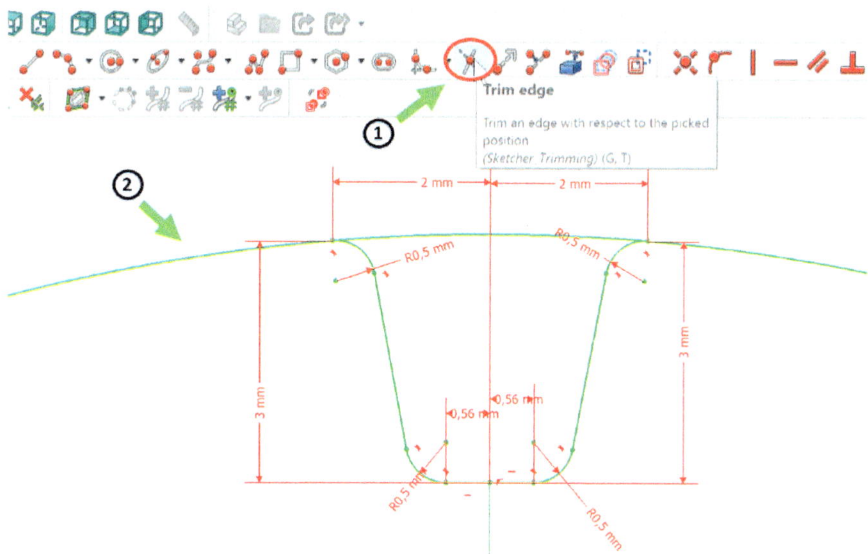

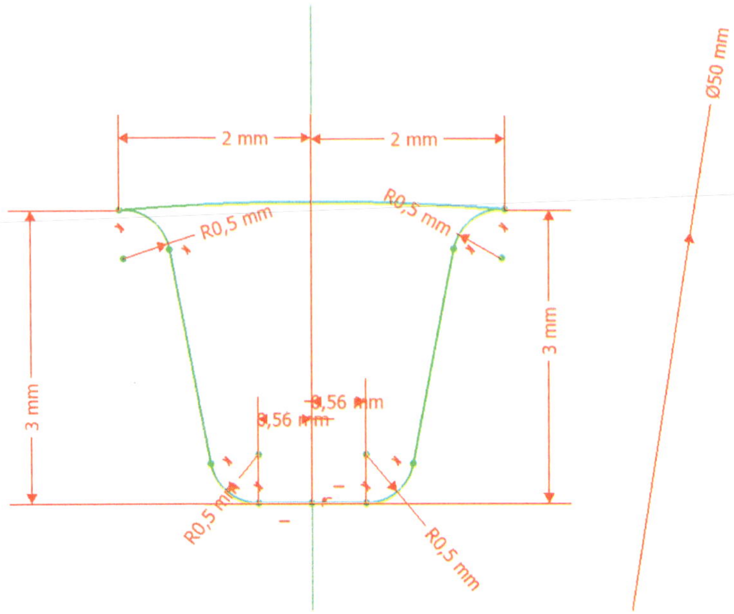

After closing the sketch, we can cut out the first tooth using the "Pocket" function. In the "Type" setting, use the "Through all" option to make a complete cutout.

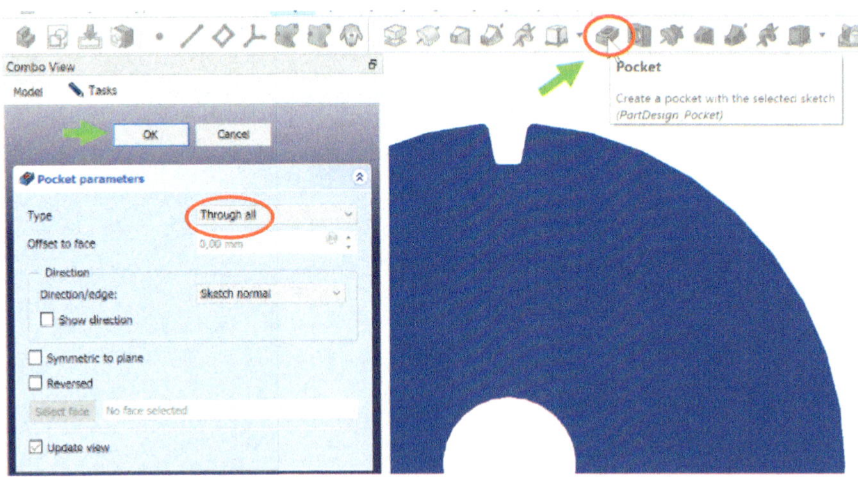

So that we do not have to create all the other teeth in this way, we will use the function "Polar Pattern" below, which allows us to create a circular pattern. This means we can use it to copy the section of the first tooth in a circular arrangement - around the center point.

To do this, we select the section we just created ("Pocket") in the Structure Tree and then click on the "Polar Pattern" command in the toolbar.

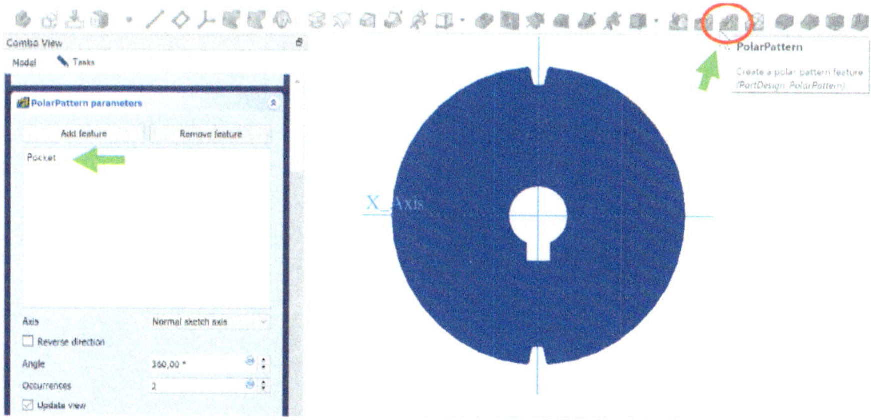

In the lower part of the command settings, we can now enter the desired number of copies. For example, we need 25 pieces. This will give us a preview of all the teeth of the gear, and with a click on "OK" we can create them.

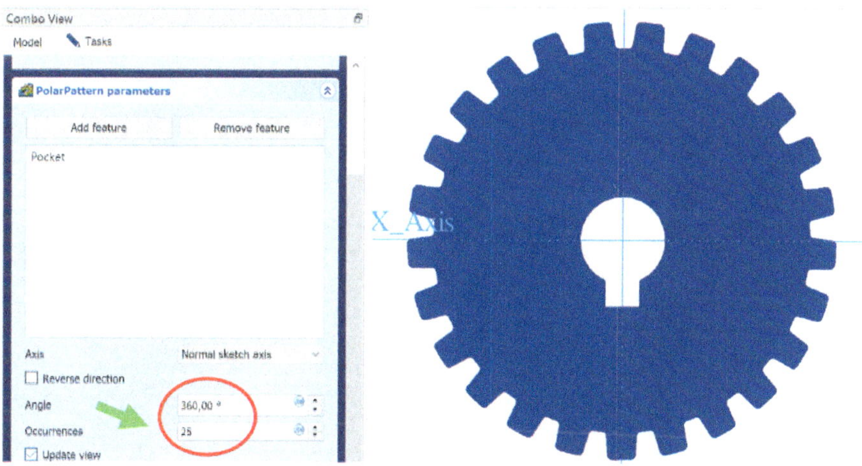

We made it! After a short break, we're off to the next project! We will create an artistic flower vase. The next section will also include a few more everyday objects, such as a remote control and a watering can.

5 Project No. 4: Flower vase

Welcome back! In this project, we will design a fancy flower vase that we will build as a simple rotary piece.

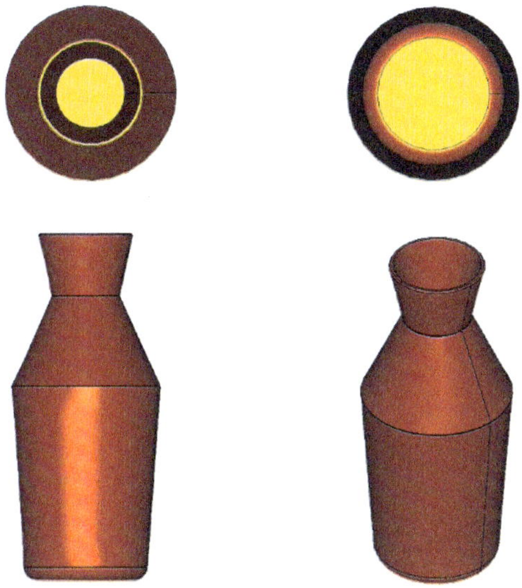

After we have created a new "Part Design" document and a body, we start a sketch, e.g., on the x-z plane and draw – as with all rotating parts – half of the cross-section of the 3D body.

We start the sketch with a 250 mm long vertical line. We set the starting point of the line to the origin of the coordinates, and we dimension the line using the constraint "Constrain vertical distance" or "Constrain distance".

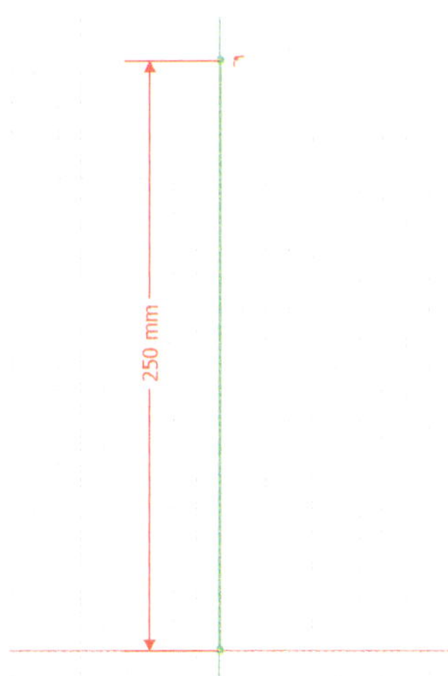

We then sketch an upper and a lower horizontal boundary line with a dimension of 35 mm for the upper line and 45 mm for the lower line.

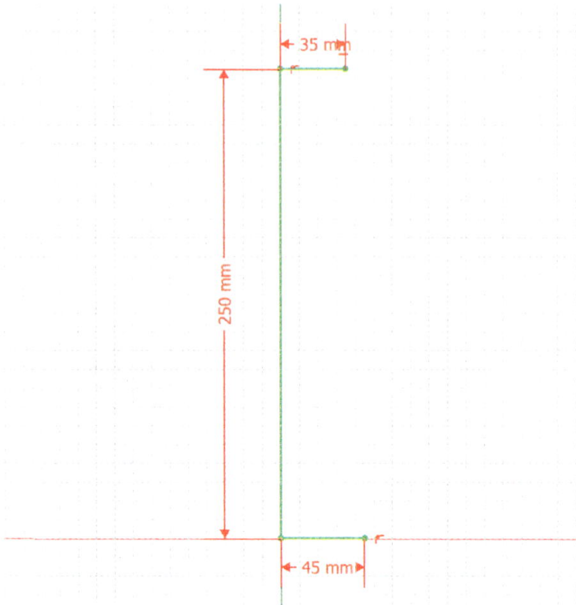

Then follow two more horizontal lines that will serve as guides for drawing the outer contour of the vase. One line should be 25 mm long and one line should be 55 mm long.

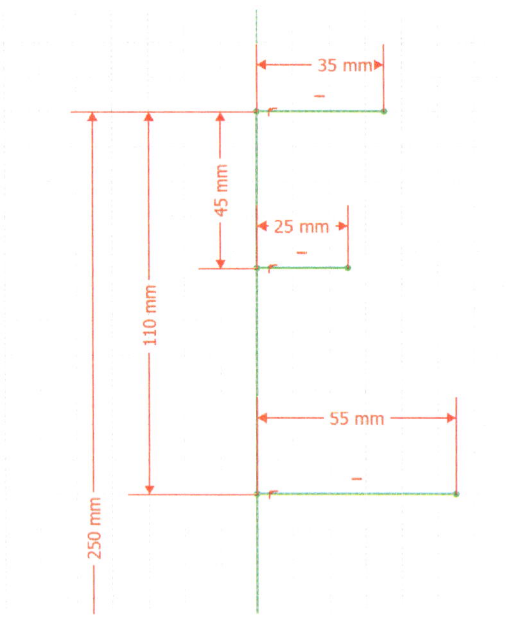

The 25 mm line is given a distance of 45 mm from the top-end point of the vertical line, and the 55 mm line is given a distance of 110 mm from the same point. Now we draw three connecting lines as shown to complete the outer contour.

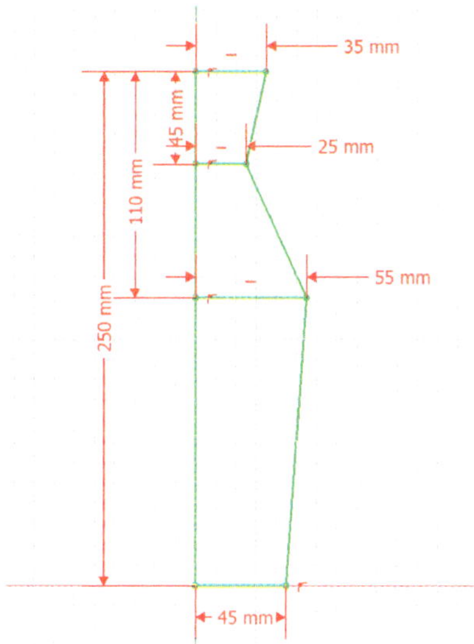

We now have to convert the two auxiliary lines created previously into construction lines. This offers the advantage that these lines are then ignored by the program in 3D mode ("Part Design"), and we thus obtain a continuous surface that we can rotate. Otherwise, we would have three individual segments. The conversion is easily done by selecting one line at a time and then clicking on the "Toggle construction geometry" command in the toolbar. The color of the line should change afterward - depending on the selection in the settings. In this case, the lines will turn light blue.

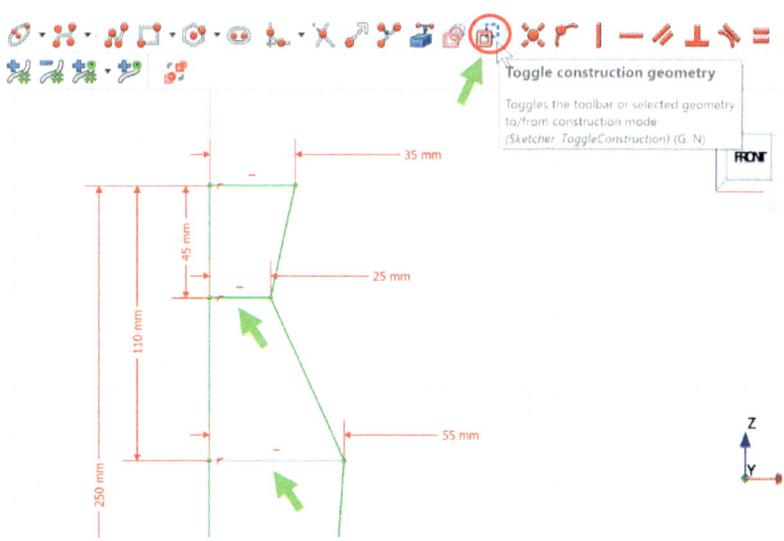

Now, we can close the 2D sketch and rotate the created half of the cross-section of the vase around an axis by clicking on the "Revolution" command.

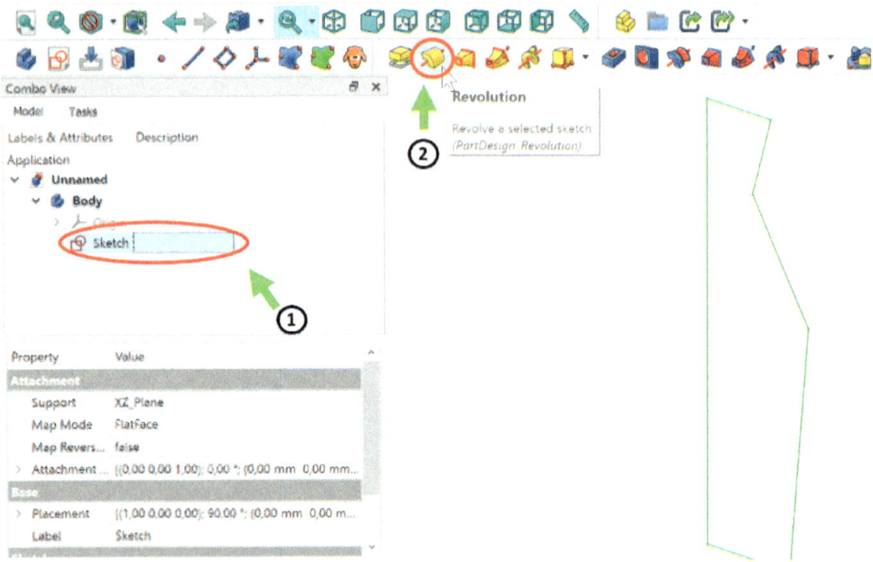

In the settings, we can leave the selection at the option "Axis" on the selection "Vertical sketch axis". The angle of 360° is also already set and just right for our model. With a click on "OK" we create the model.

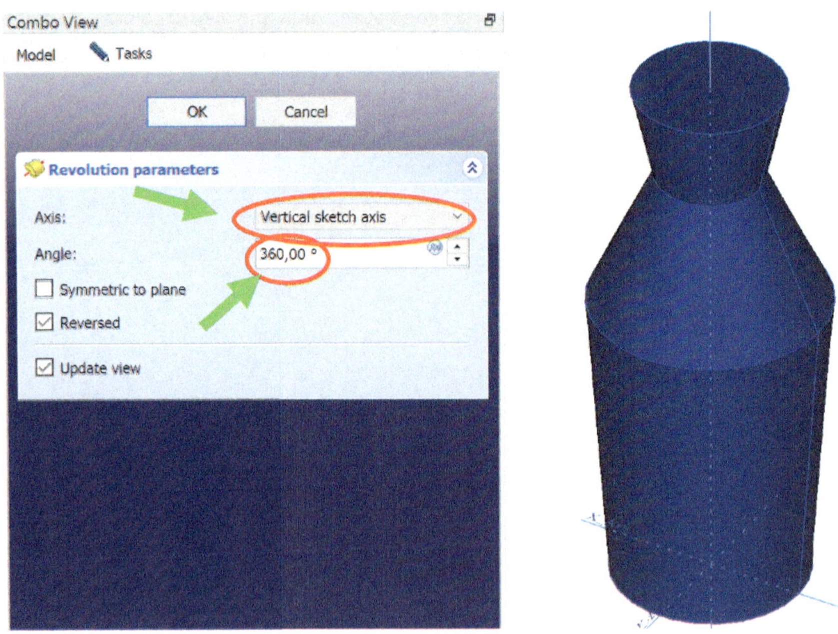

The basic body of the flower vase is now ready. Next, we hollow out the body by clicking on the top surface of the vase and using the "Thickness" command.

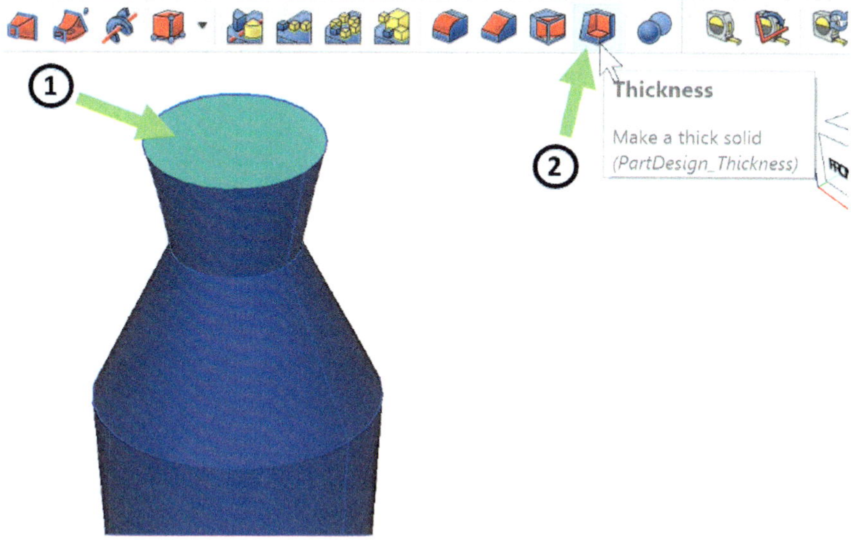

For example, for the wall thickness, we can choose a value of 3 mm. It is important that we activate the option "Make thickness inwards" so that the outer dimensions of the vase do not change.

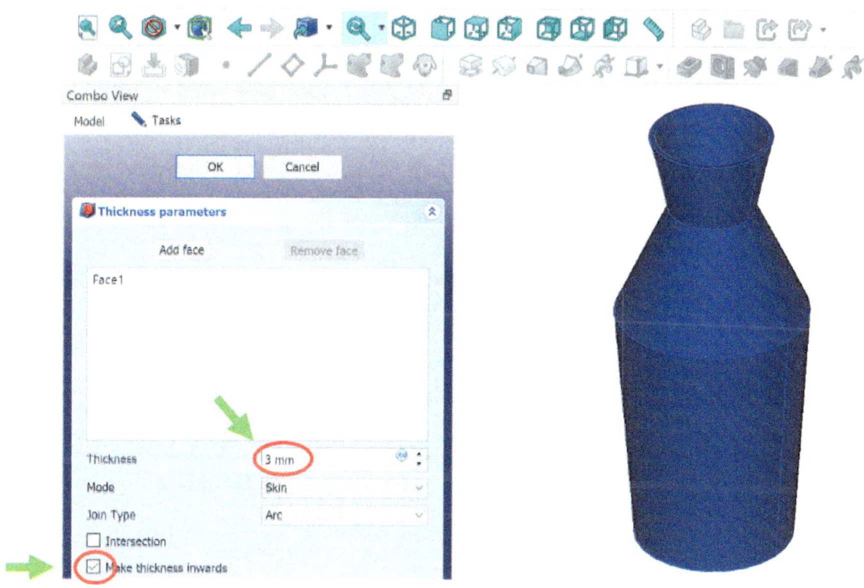

To soften the angular design a bit more, we add a 10 mm fillet for the bottom edge.

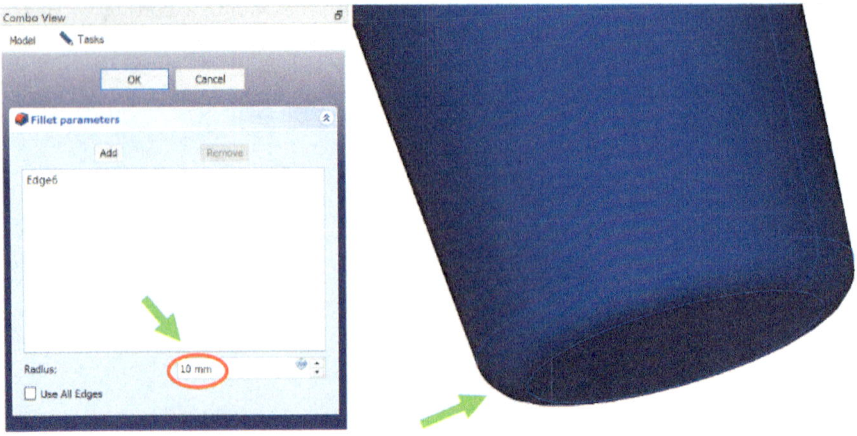

For the three remaining edges, for example, we select fillets of 1 mm. Multiple selection is possible by holding down the CTRL key.

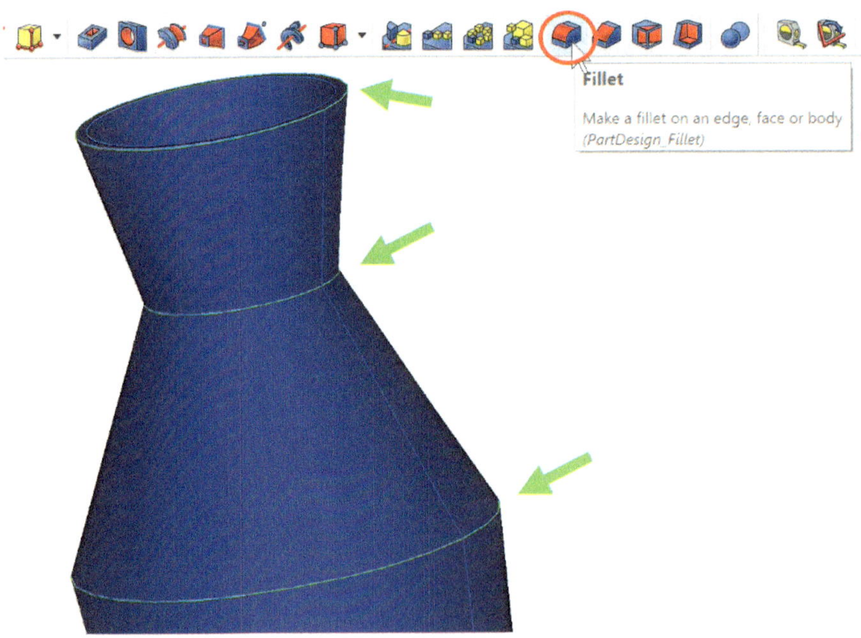

Finally, we would like to change the appearance of the flower vase a bit. By right-clicking on the body in the structure tree and selecting "Appearance ..." we can adjust the appearance to our liking.

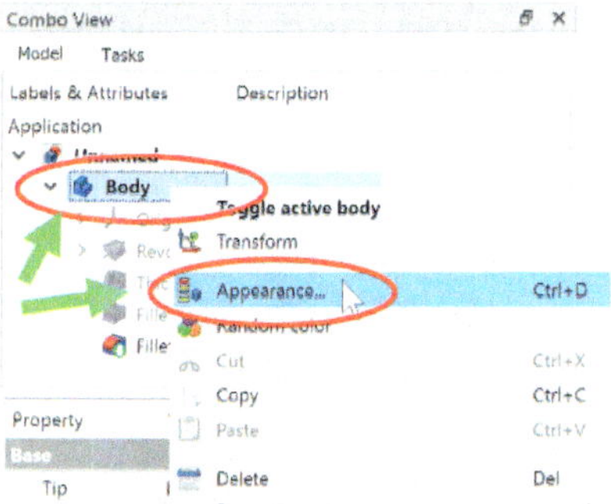

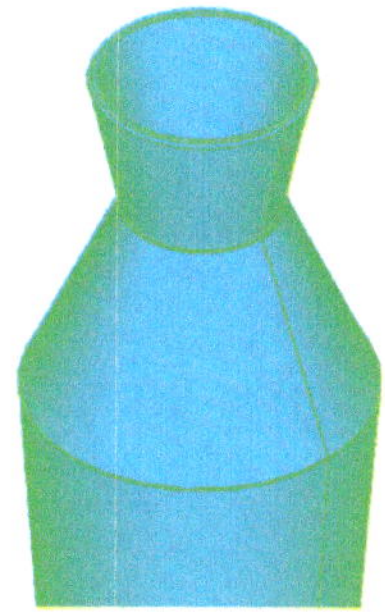

Here we can change the color of the body on the one hand, but on the other hand we can also change the material of the body here. If we change the material of the body, the typical appearance of the selected material will also be transferred to the 3D model. For example, we could select the material "Copper" and thus obtain a copper-colored vase.

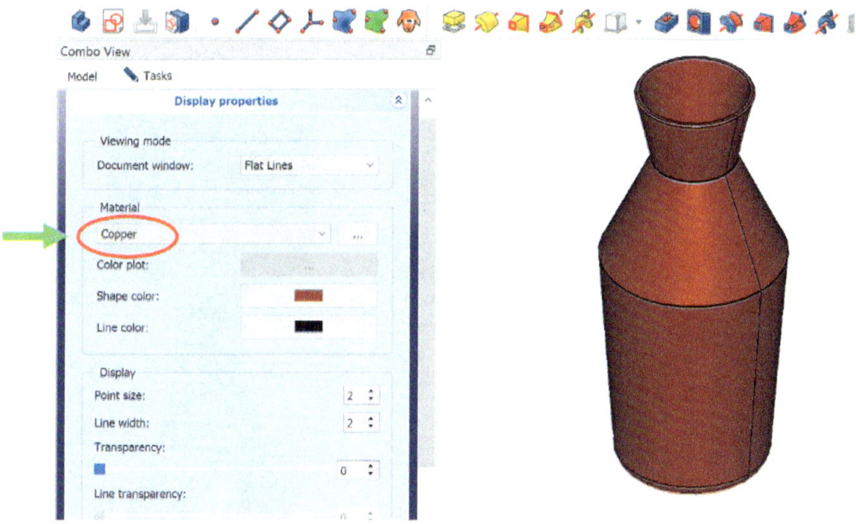

Very good! In the next two projects, we will once again devote ourselves to somewhat more difficult designs. We will design a set of pistons, connecting rods and piston pins as well as a wrench. Then we will deal with the design of a ball bearing, as well as a remote control and a watering can.

6 Project No. 5: Connecting rod, piston, and piston pin

In this project, we would like to design a set of piston, connecting rod and piston pin.

To do this, we start with the creation of the flask. For this, we start a new document in the workspace "Part Design" and create a body ("Create Body"). For the basic shape of the piston, we need a cylindrical body. We create this by extruding a sketch that we draw on the x-y plane. In this sketch, we draw a circle with 85 mm diameter, and we can close the sketch afterward.

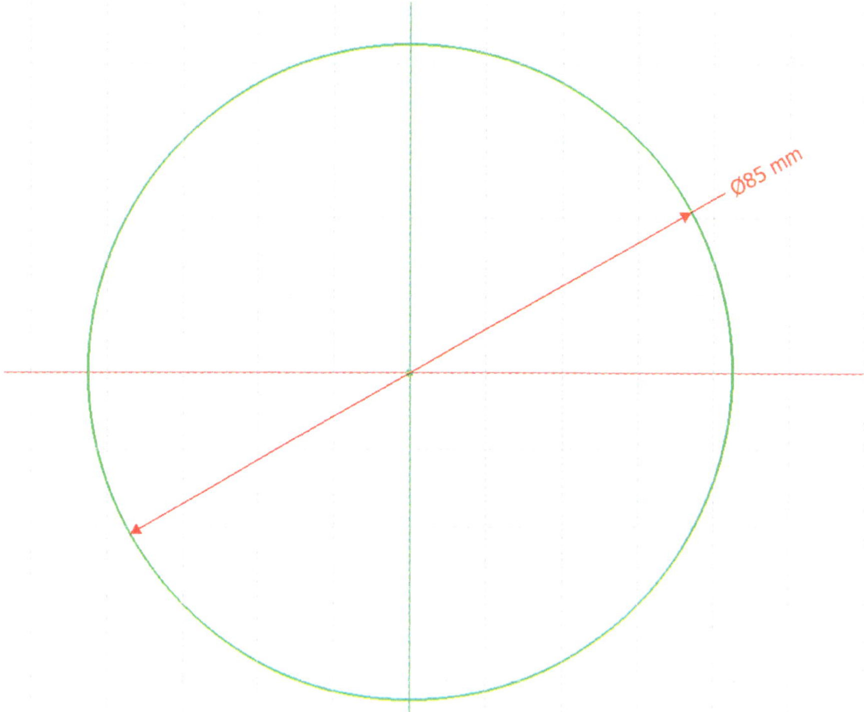

Now we need to extrude the circular area using the "Pad" 70 mm command.

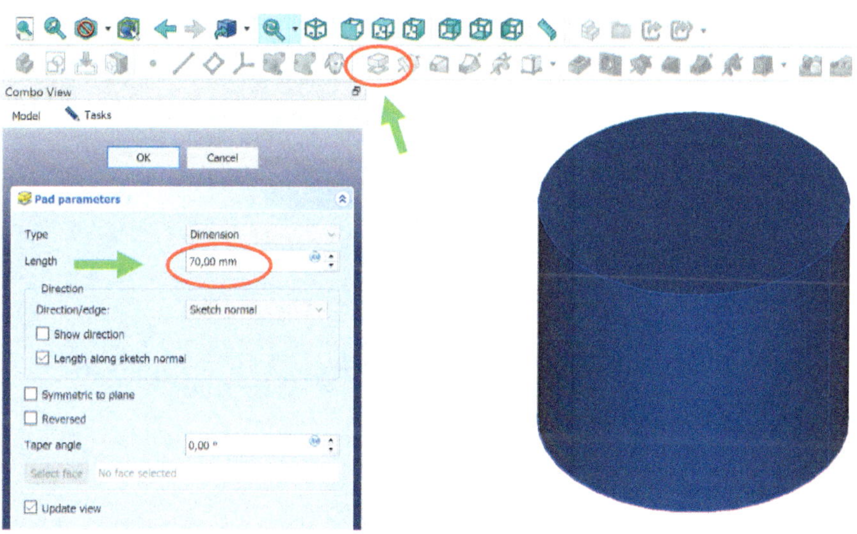

In the next step, we hollow out the piston by selecting the bottom face of the model and then clicking on the "Thickness" button in the toolbar.

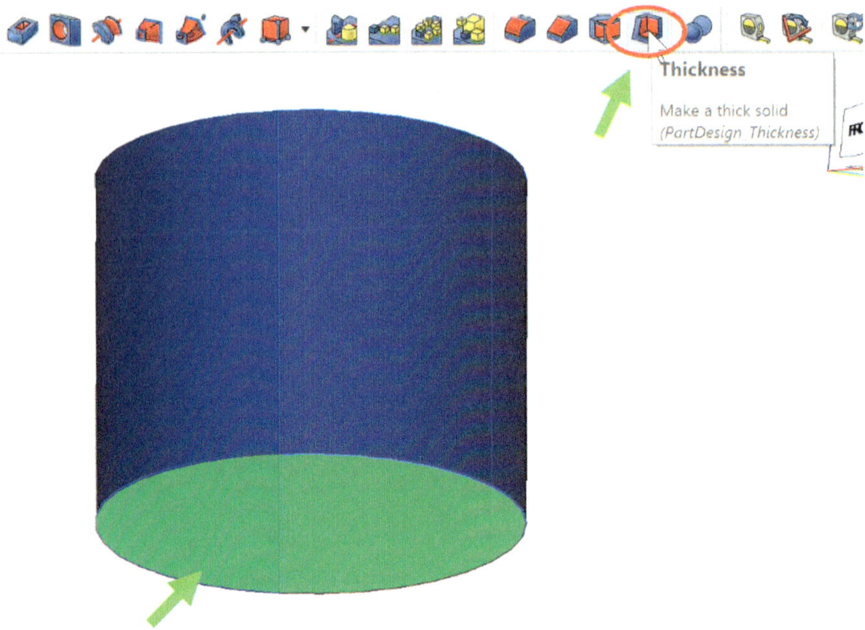

We assign a wall thickness of 5 mm ("Make thickness inwards" activate).

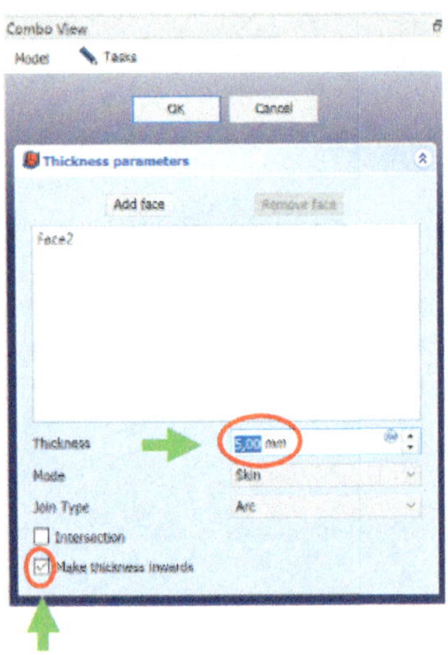

Then we start a sketch on the x-z plane to make a cutout for the piston pin, which will later connect the piston to the connecting rod. After hiding the body using the space bar, we draw a circle with a diameter of 30 mm on the green vertical axis. Then we dimension the circle with 30 mm distance from the coordinate origin.

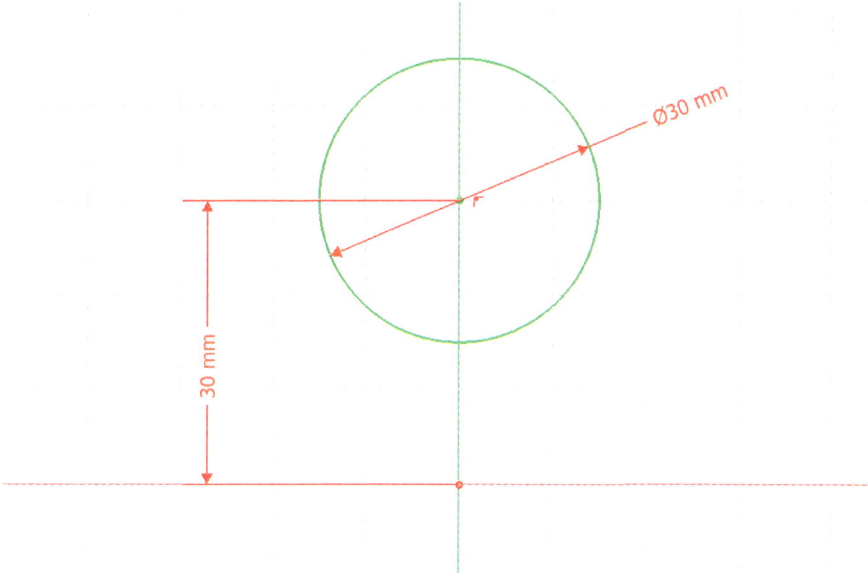

After that, we can close the sketch and show the body again. We then select the sketch we have just created in the structure tree and then click on the "Pocket" command to create a section through the complete part.

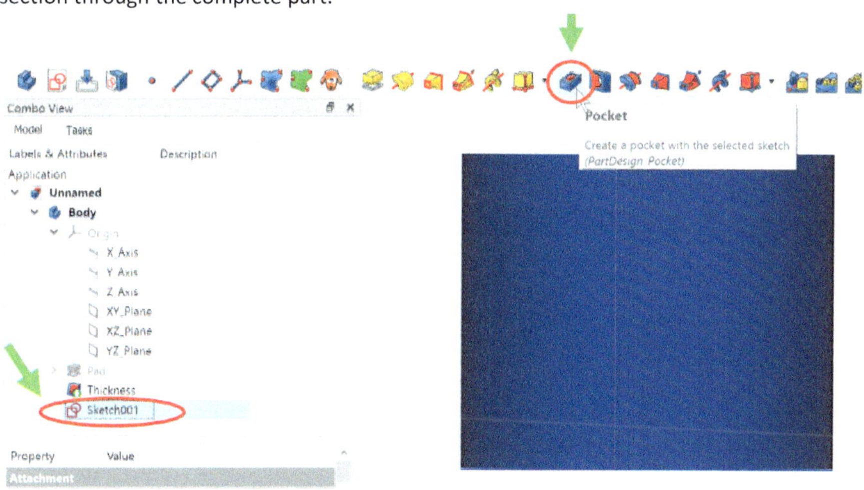

So that this cutout goes in both directions through the complete part, we select in the setting "Type" the option "Two dimensions" and enter in the options "Length" and "2nd length" 50 mm each.

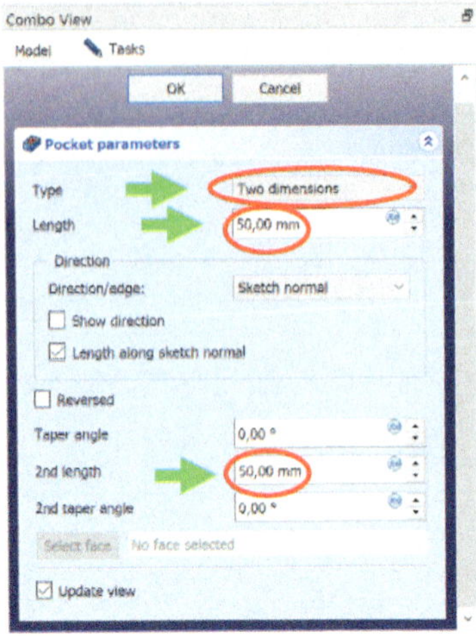

Next, we add two grooves that will be provided for mounting two piston rings. We do this with the command "Groove". For this command, we need a sketch with a profile, which we will rotate around an axis to remove material. We create the sketch on the x-z plane. We hide the body again with the space bar for better clarity.

We create two identical 2 mm x 3 mm rectangles with a distance of 5 mm between them. The rectangles should lie in a line, so we set a constraint between two vertices of the rectangles using the command "Constrain vertically".

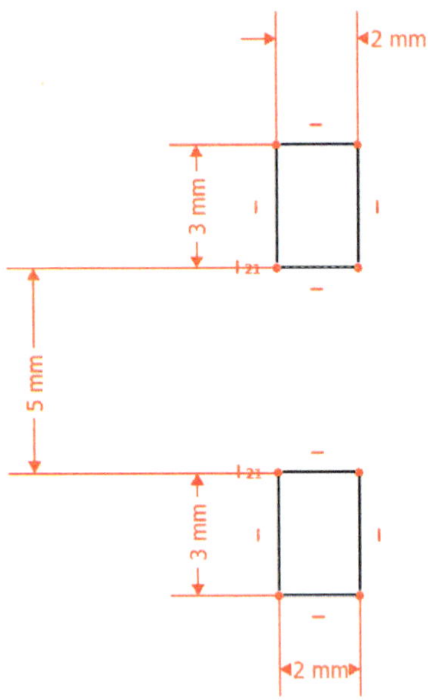

We then dimension the lower-left corner of the lower rectangle at 40.5 mm (horizontal) or 52 mm (vertical) to the coordinate origin.

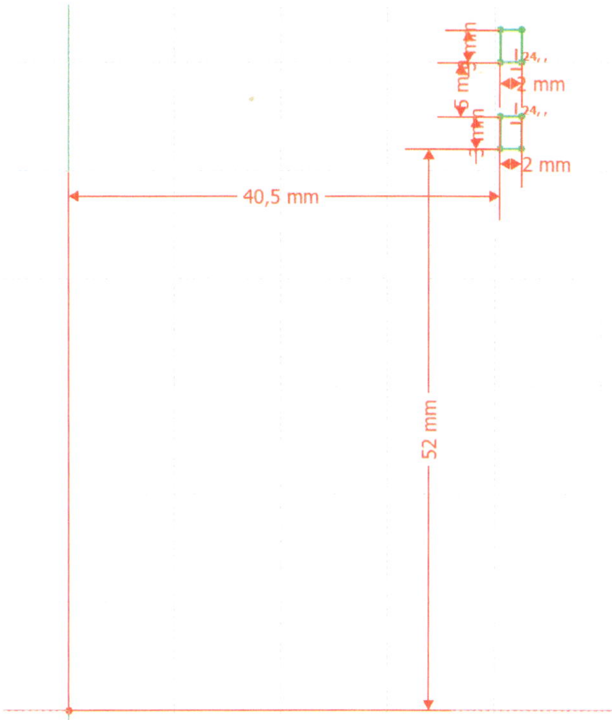

Now we can close the sketch and show the body again. After that, we make sure that the sketch is selected in the structure tree and click on the command "Groove".

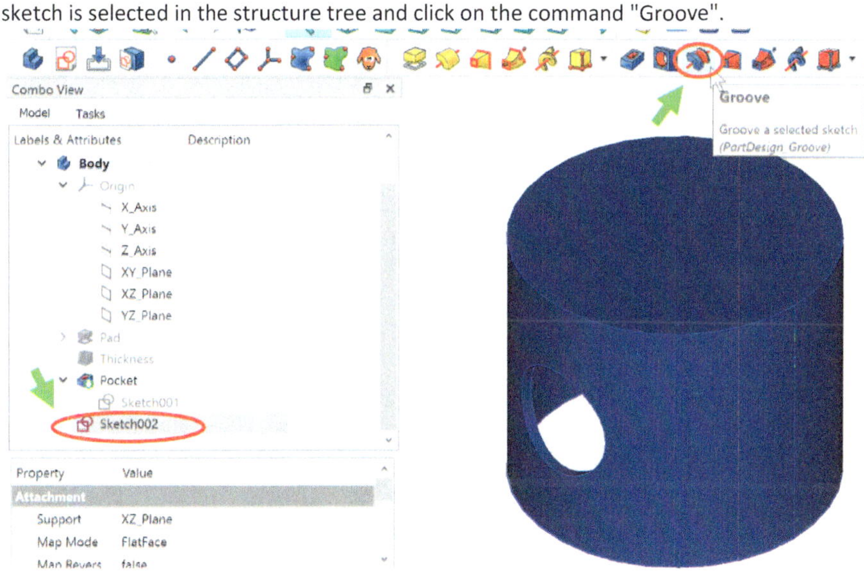

The program cuts the sketched profile out of the body in a rotating motion around the z-axis, giving us the desired rectangular grooves.

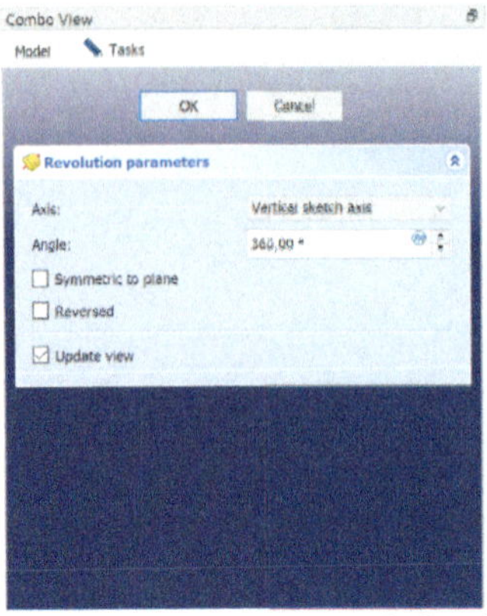

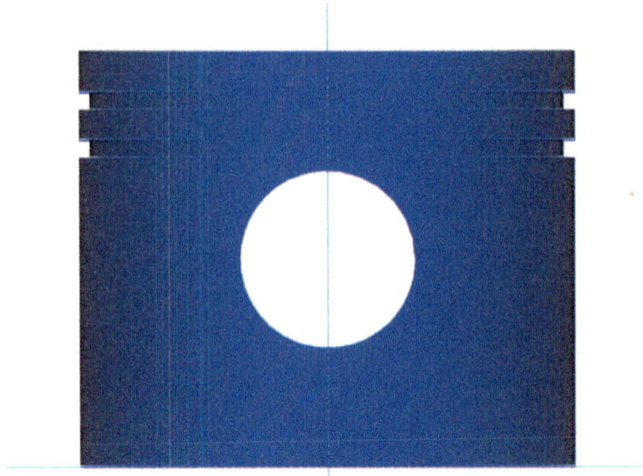

As a further detail, we similarly add a trough in the upper area of the piston, which normally ensures optimum mixture and pressure distribution. To do this, we create a new sketch on the x-z plane and draw the following triangular profile. One point of the triangle should be connected to the green vertical axis.

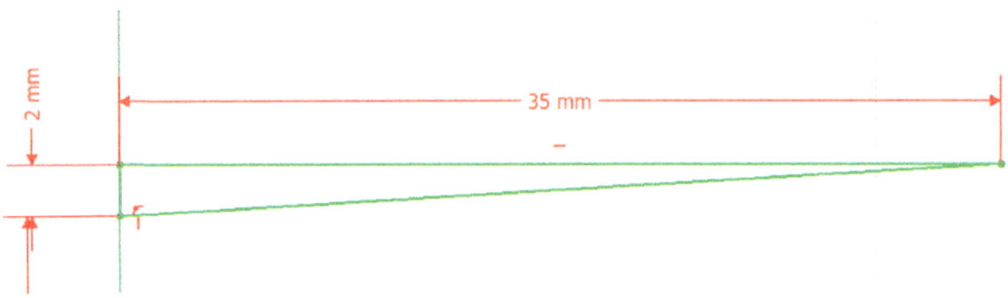

We dimensioned this profile – starting from the lower-left corner – with a distance of 68 mm to the coordinate origin.

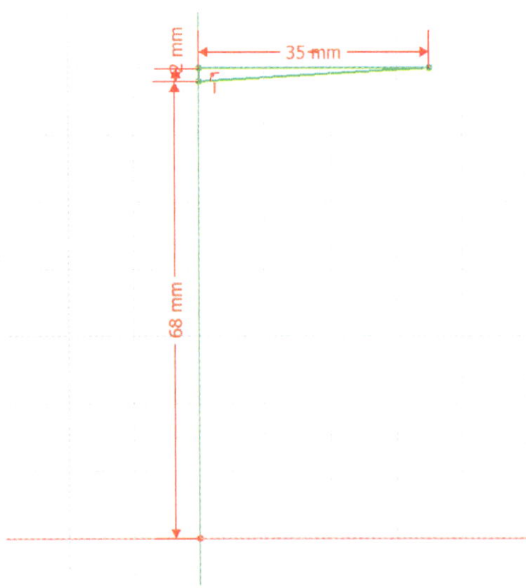

Then we can close the sketch.

After we have blended in the body and selected the sketched profile, we again use the "Groove" command, which creates the desired trough.

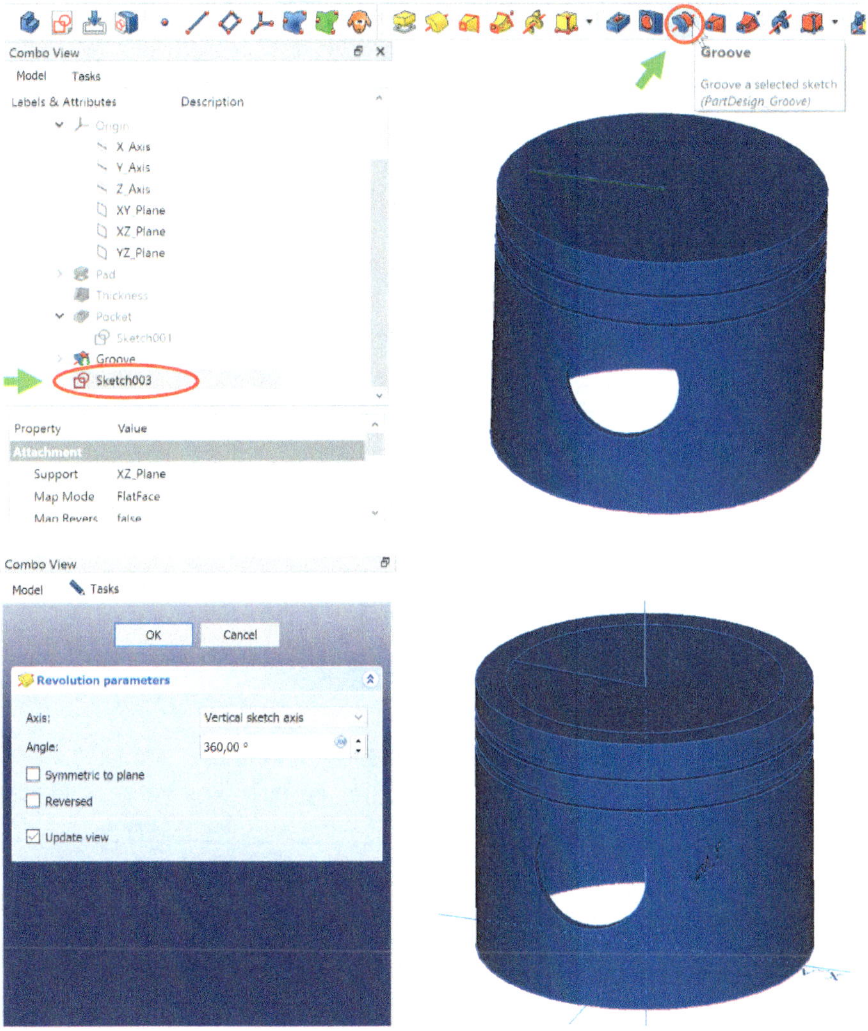

In the penultimate step, we round off the upper and lower edges of the piston with 1 mm each.

Finally, we create two cutouts at the openings for the piston pin.
To do this, we first create a plane parallel to the x-z plane with a distance of 42.5 mm. We do this with the command "Create a datum plane".

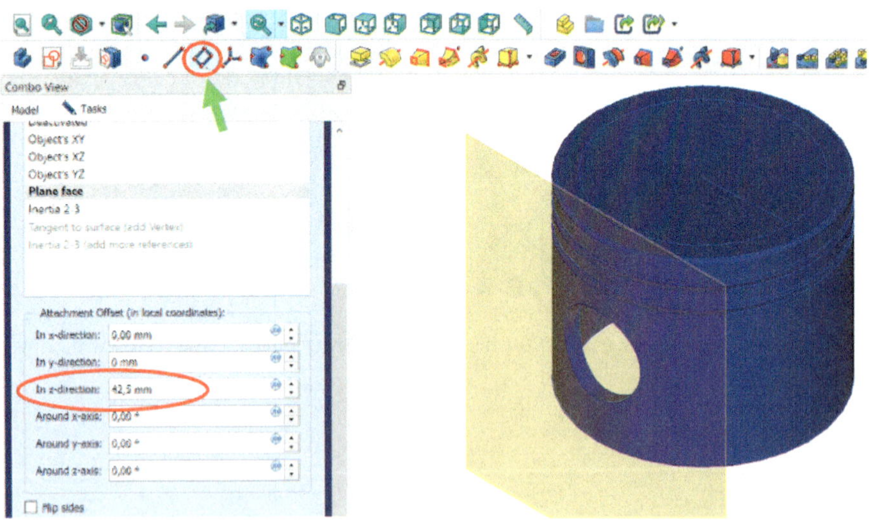

On this layer, we create a new sketch in which we draw a 45 mm wide and 30 mm long rectangle whose center ("Create a centered rectangle" use) lies on the green vertical axis. We dimension the distance from the center of the rectangle to the coordinate origin as 30 mm.

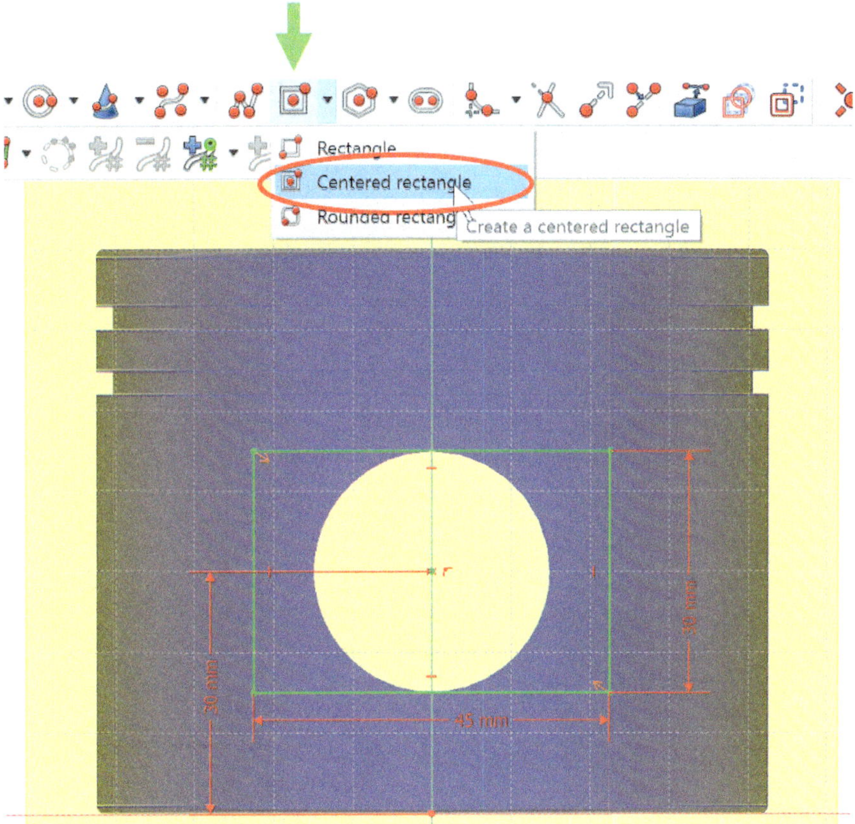

Based on this sketch, we can then create a cutout with a dimension of 3 mm using the "Pocket" command.

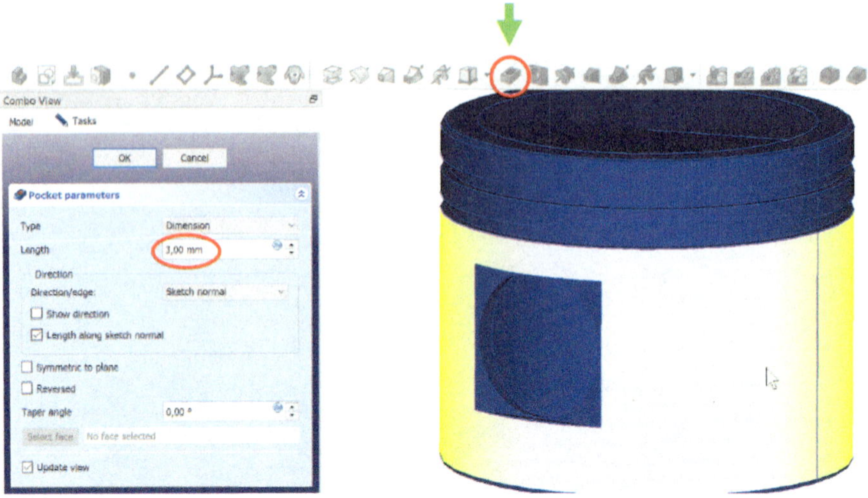

We need the same section on the opposite side. We create it by simply mirroring the first section. For this, we use the command "Mirrored". We select the x-z plane as the mirror plane.

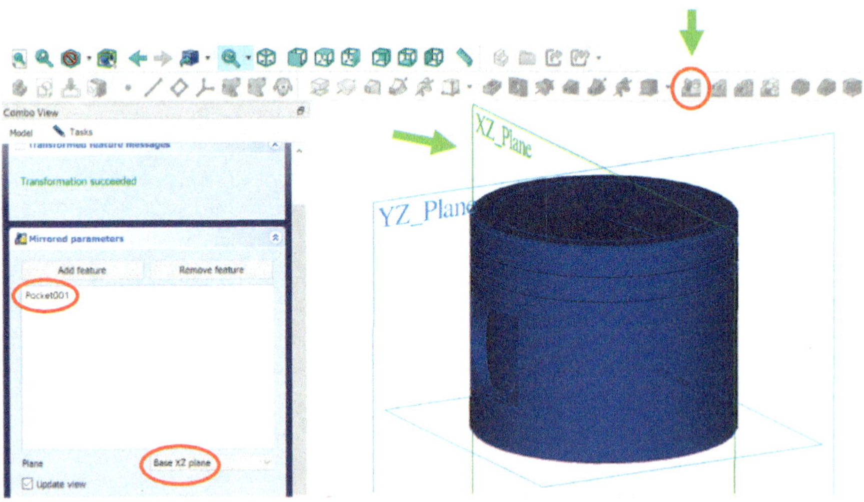

We will omit further details for reasons of complexity and time. This piston will only represent a model anyway.

We now continue with the connecting rod and the piston pin. But before that, we have to save the piston. Then we can close the document.

For the connecting rod we again create a new document, a body, and a sketch on the x-z plane. We sketch on this plane the cross-sectional profile of the connecting rod shown.

We start with the two "eyes" for this. The upper connecting rod eye should have a diameter of 30 mm (inside) and 40 mm (outside). We set the circle centers to the coordinate origin.

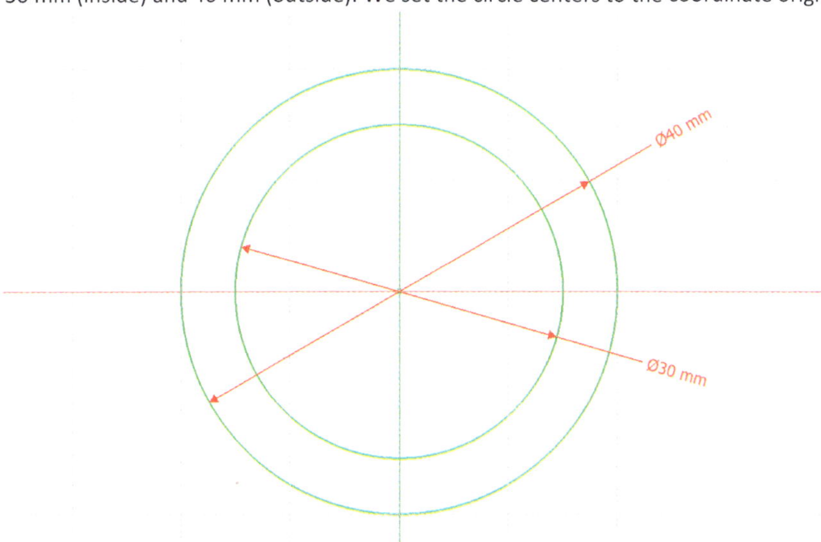

A little further down, we draw the second connecting rod eye, which also consists of two circles whose centers are congruent and should lie on the green vertical line. The circles should have a diameter of 50 mm (inside) and 80 mm (outside). In addition, we dimensioned the distance between the centers and the coordinate origin as 165 mm.

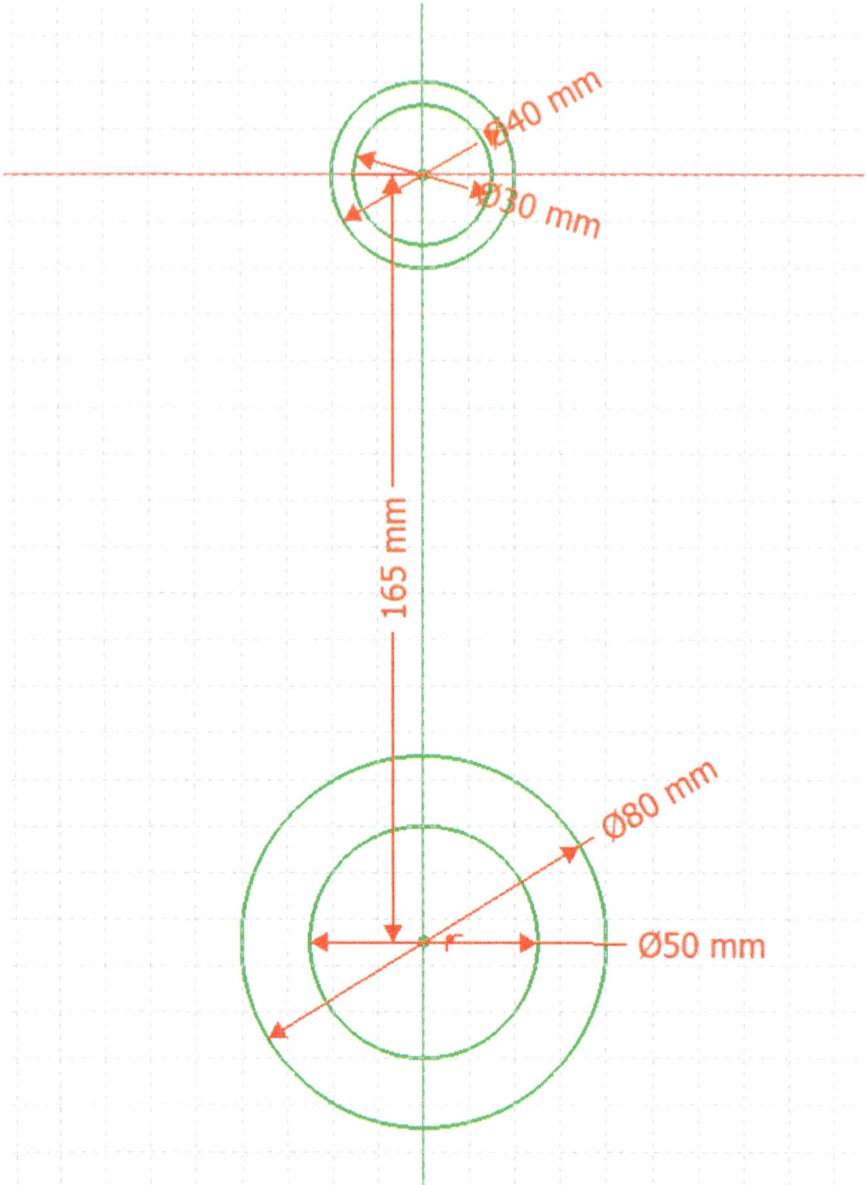

Then we draw two 65 mm long vertical lines, each of which should have a horizontal distance of 10 mm from the center of the upper connecting rod eye.

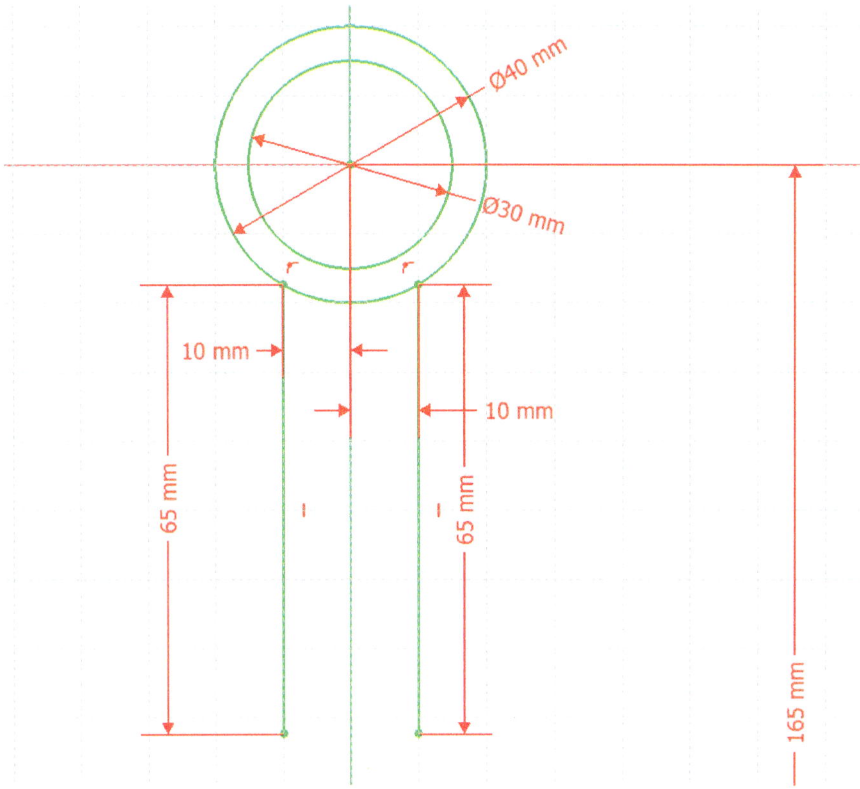

We complete the profile with two arcs ("End points and rim point"), which are to receive tangential links ("Constrain tangent") at the end points in each case.

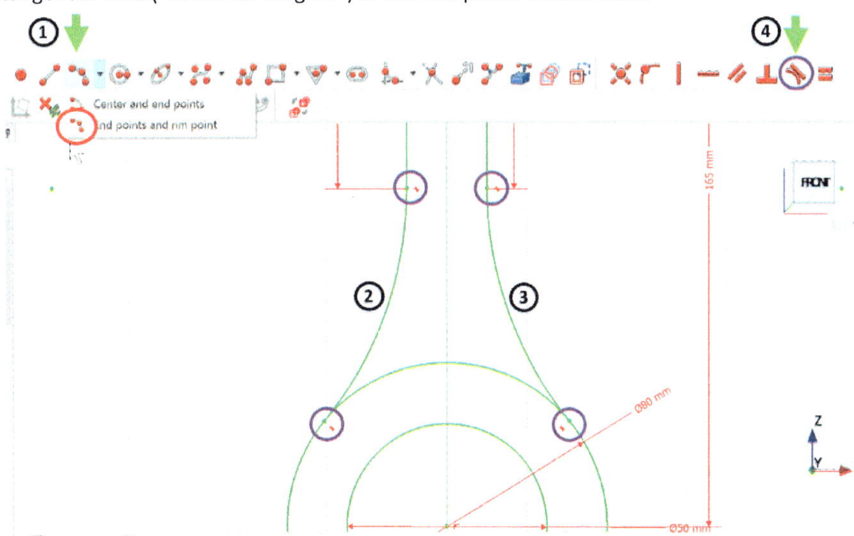

Finally, we use the function "Trim edge" to remove excess sections.

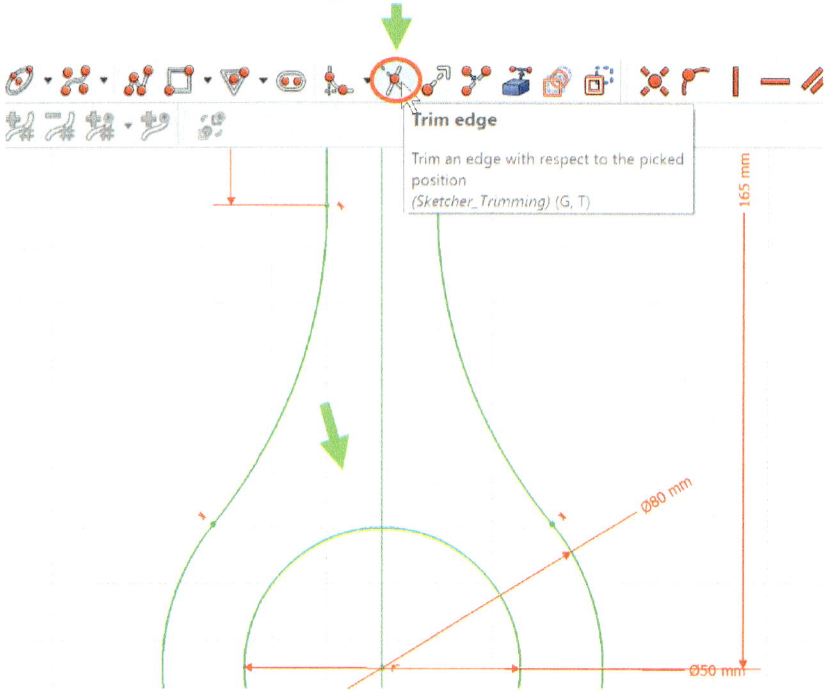

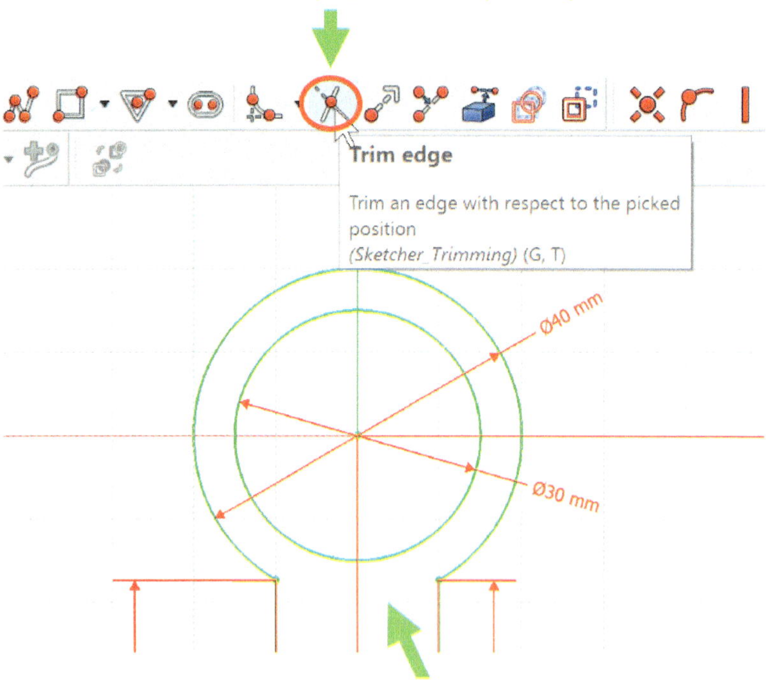

When this is done, we can close the sketch and extrude the connecting rod 20 mm using the "Pad" command.

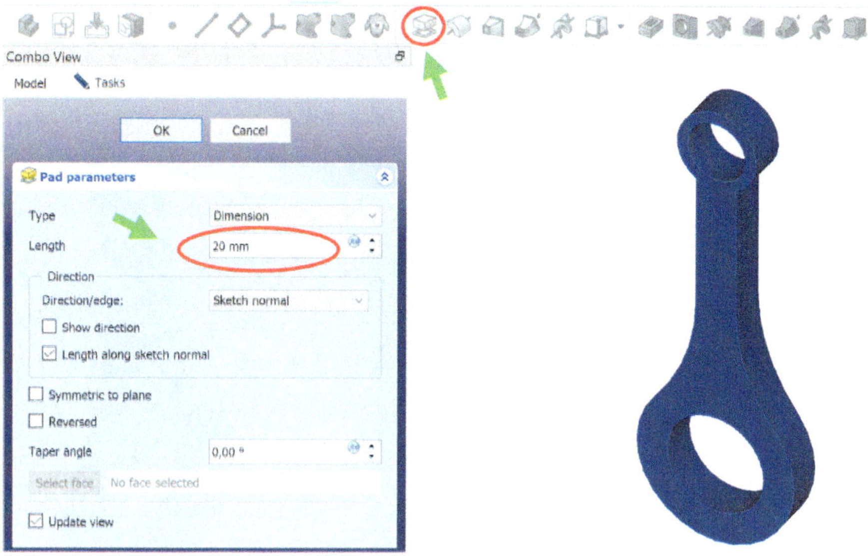

So that the transitions in the upper area of the connecting rod are not too extreme, we can round off these edges with a radius of 20 mm.

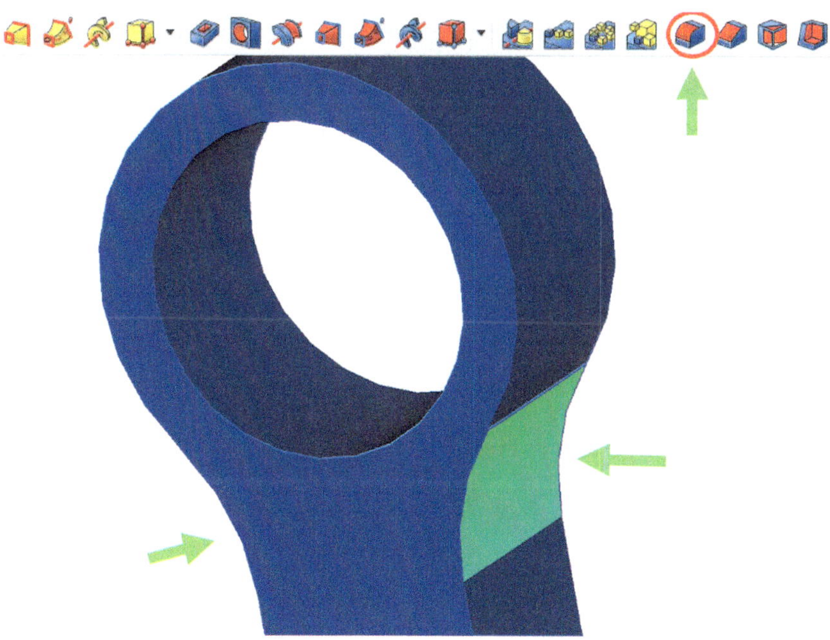

We also round all the remaining edges of the front and back sides by 1 mm each. To do this, we select the front and back sides with the CTRL key pressed and then click on the command "Fillet".

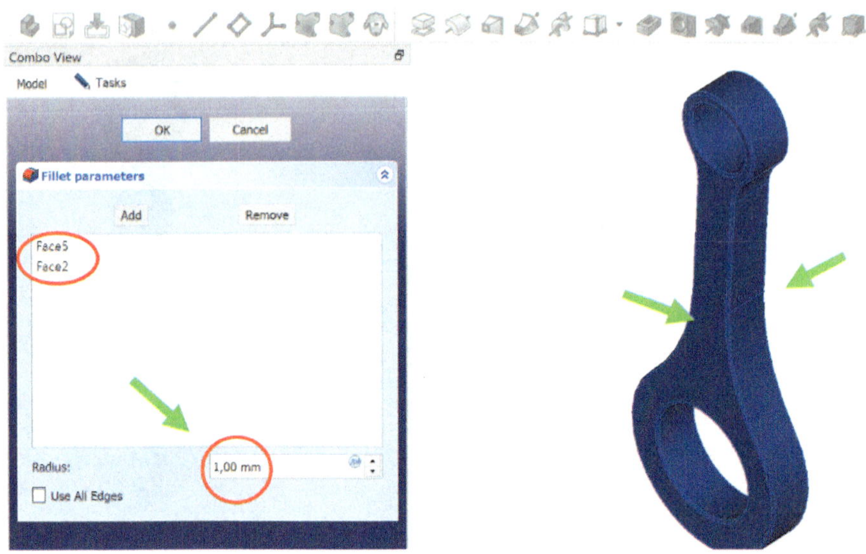

Our 3D body in this case is a highly simplified model.

Normally, the lower area of the component is divided into two parts, the geometry is more functional and, in addition, there are the so-called connecting rod bearing shells, which would sit in the lower eye and serve as plain bearings.

We can then save and close the connecting rod. We then draw the piston pin before virtually assembling the set of piston, connecting rod and pin. For this, we create a new document, a new body and draw a circle with a diameter of 30 mm on its y-z plane, which we then extrude 79 mm and hollow out to a wall thickness of 3 mm. You are welcome to do this independently for better practice.

Here is the solution:

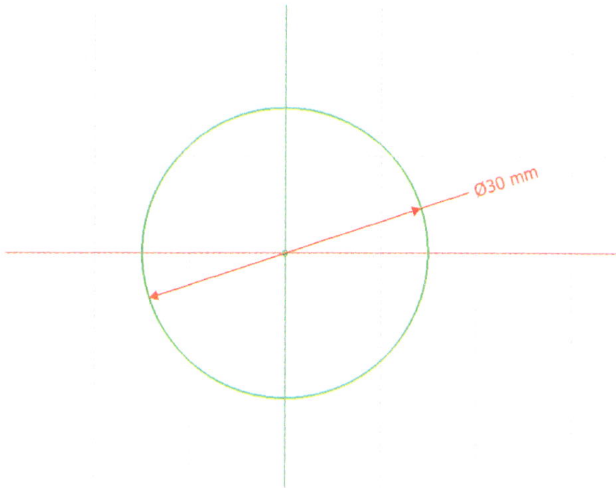

For the "Thickness" command, we select both faces of the stud. In addition, we activate the option "Make thickness inwards".

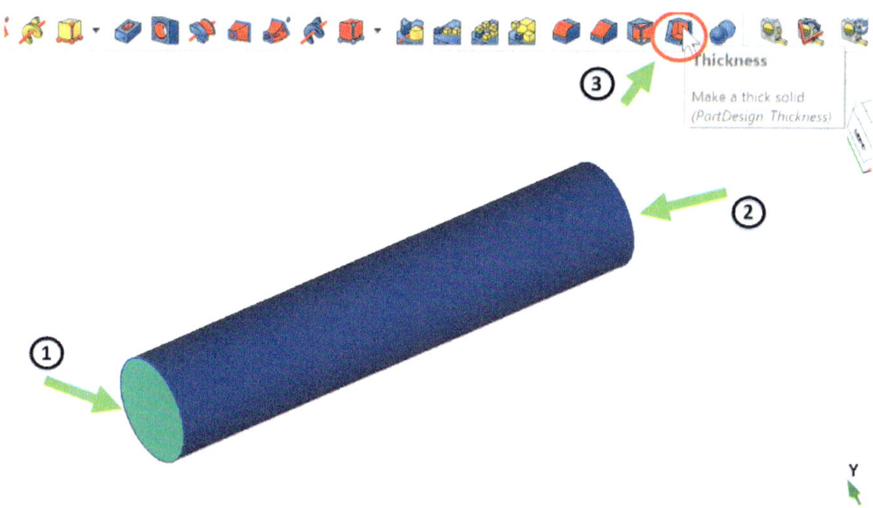

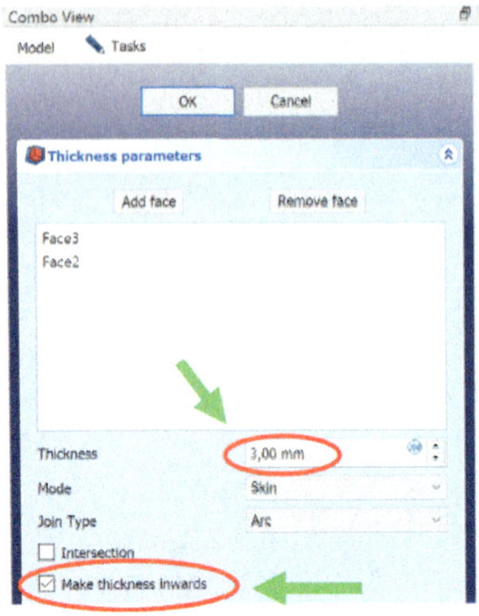

Finally, we save and close this 3D model as well.

In the following, we would like to assemble the components virtually. We do this in the workspace "A2plus". However, this area is not installed by default. You can install the area manually via the Add-on Manager if required.

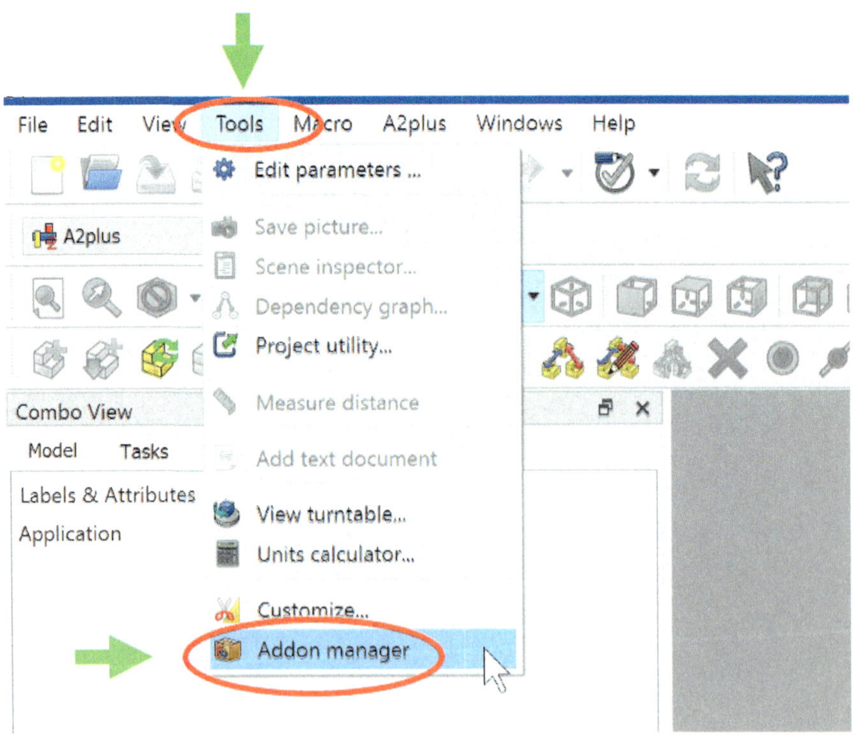

We are now in the workspace "A2plus". Here we must first create a new document, as usual. Then we can add the first 3D body, the connecting rod, to our assembly. We do this by clicking on the command "Add a part from an external file" in the toolbar. Before that, we need to save the document.

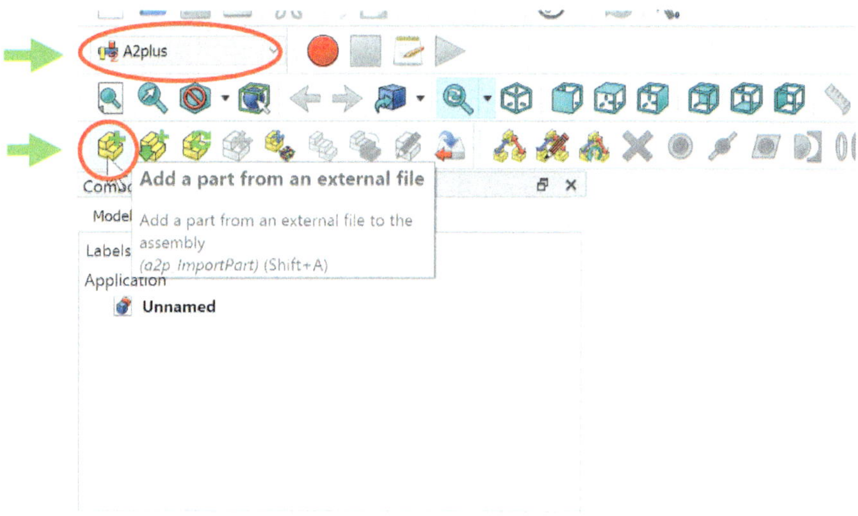

The first component that we insert into an assembly is fixed so that we can mount all other components to it. The second component we insert in the same way is the piston pin. We place it with one click at any position in the 3D space.

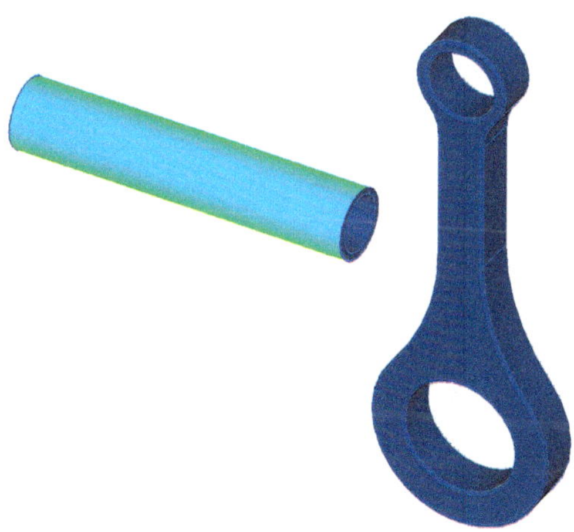

Then we select the shell surface of the cylindrical piston pin as well as the inner surface of the upper connecting rod eye (pressed CTRL key) and connect the two 3D bodies using the command "Add axis Coincident constraint".

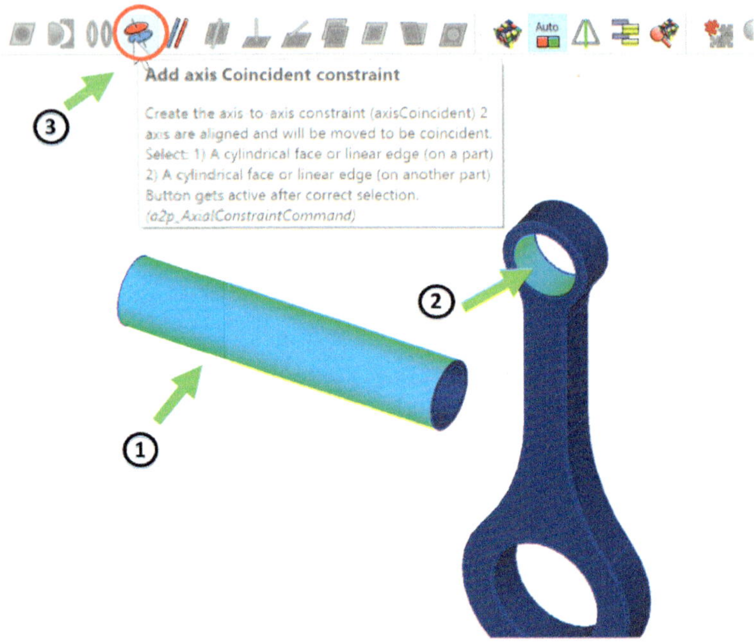

We have thus achieved that the axes of the two elements are aligned with each other. We confirm with "Accept".

Now the bolt is only movable along this axis. We can check this with the command "Move the selected part under constraints".

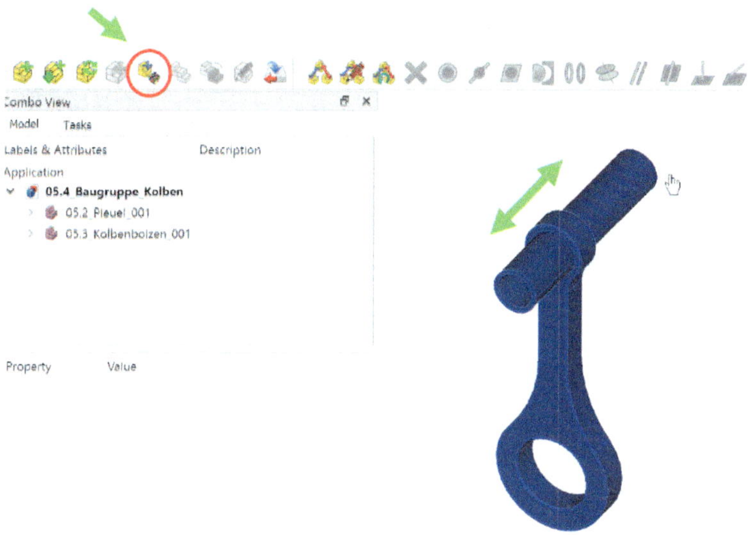

To ensure that the piston pin is exactly centered, we add a "planeCoincident constraint" condition by selecting the two front faces of the 3D bodies and the command.

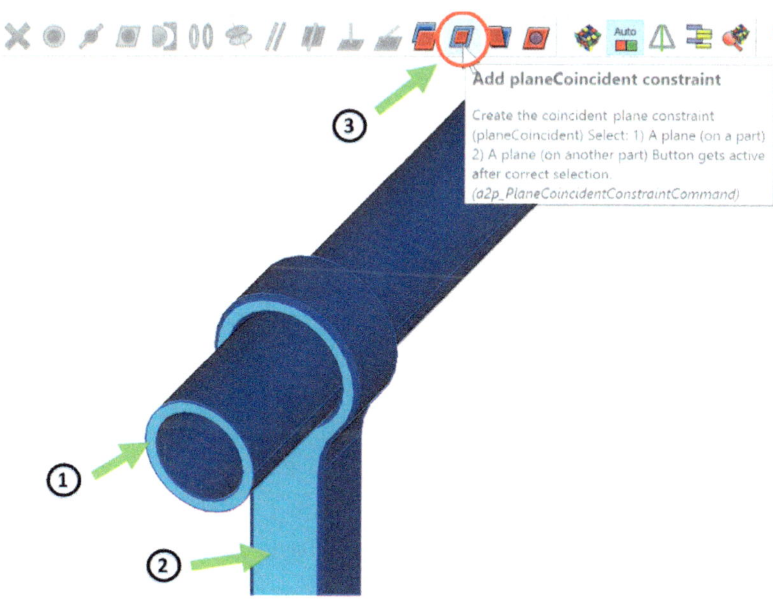

In the settings (pop-up window), we then enter an offset of 29.5 mm so that the bolt is exactly centered. Confirm with "Accept".

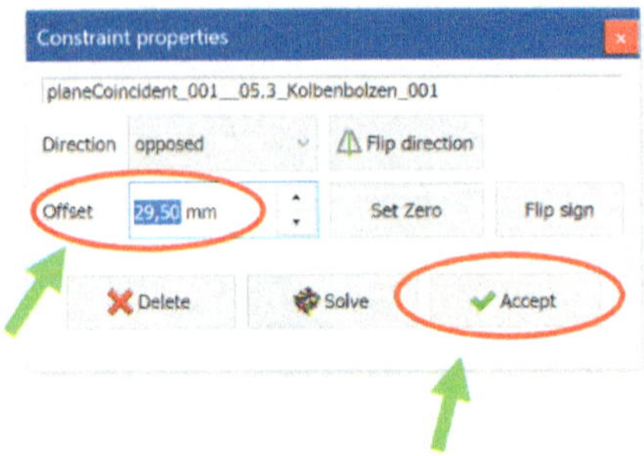

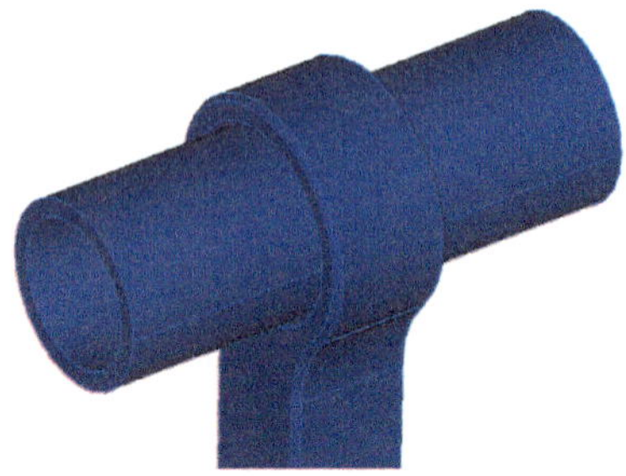

Finally, we add the piston to the assembly. You are welcome to link this to the piston pin independently.

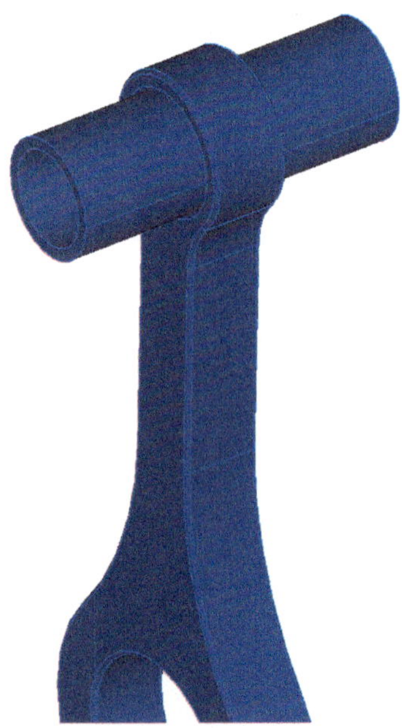

For the link, we first execute the command "Add axis coincident constraint" again.

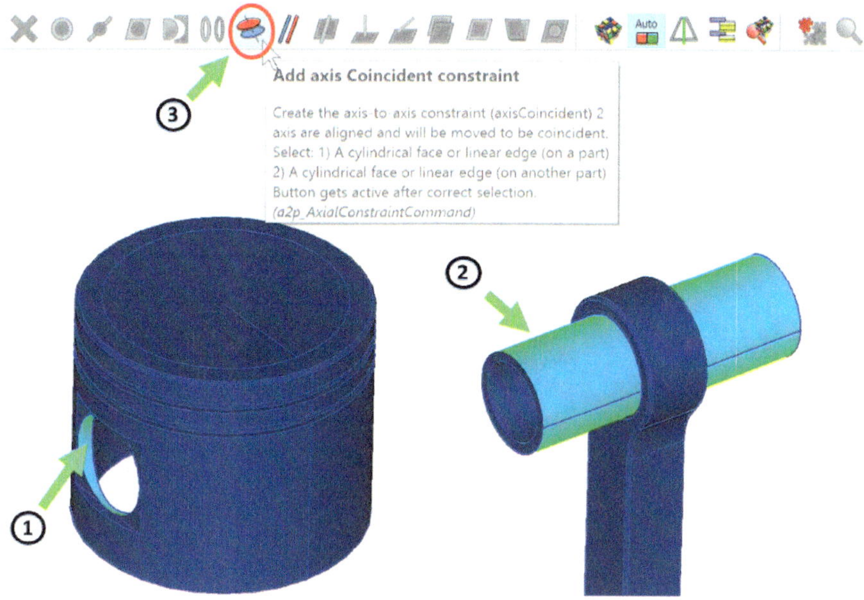

After that, we link the two displayed surfaces using the command "Add planeCoincident constraint".

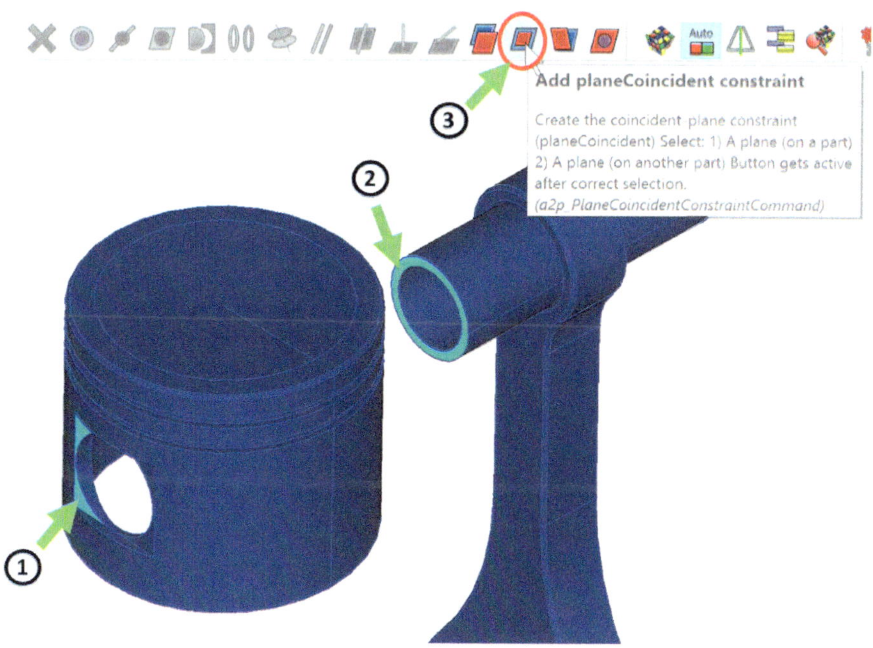

Now the assembly is ready. To better distinguish the individual parts from each other, we can change their material or color. We have already learned how this works in the last project. You can also make these changes directly in the assembly. However, the changes are then only made for the assembly and are no longer available when you open the individual parts separately. In this case, you can change the materials or colors according to your wishes.

Fittingly, the next project continues with a wrench. Be sure to stick around, there are more great and even more complex design projects in the next section, including, for example, a ball bearing.

7 Project No. 6: Wrench (open-end wrench)

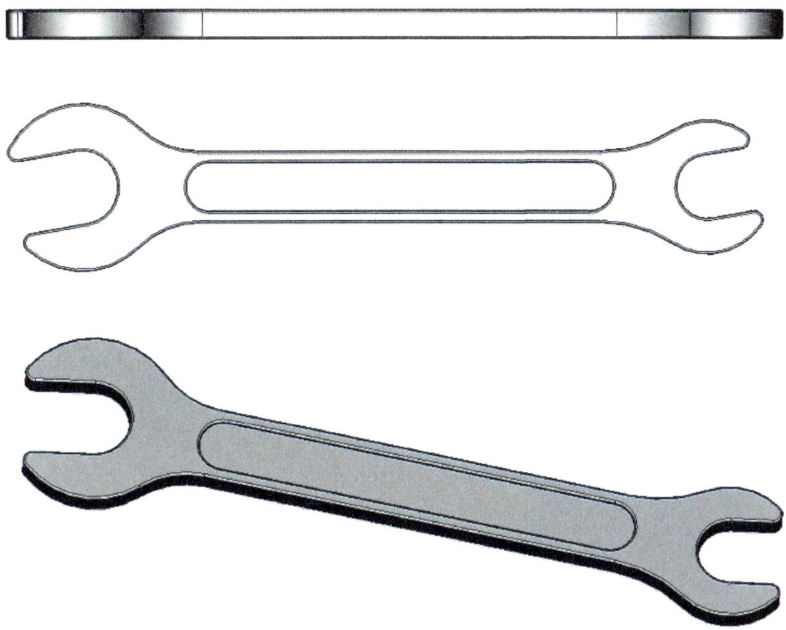

How can we best design this wrench? If we take a closer look at the geometry of the wrench, one or the other may already recognize that it makes sense to start with a circular geometry in the left and right area. Subsequently we will design the middle area of the wrench using arcs and connecting lines. The other details will follow later.

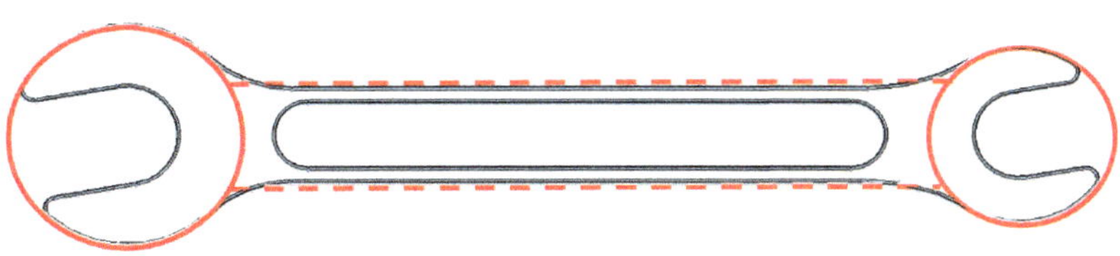

So, let's first sketch two circles on the x-y plane. The left circle should have a diameter of 35 mm and the right circle a diameter of 28 mm. We give the left circle a distance of 67 mm from the origin and the right circle a distance of 65 mm. Make sure that the centers of each circle are on the horizontal red axis.

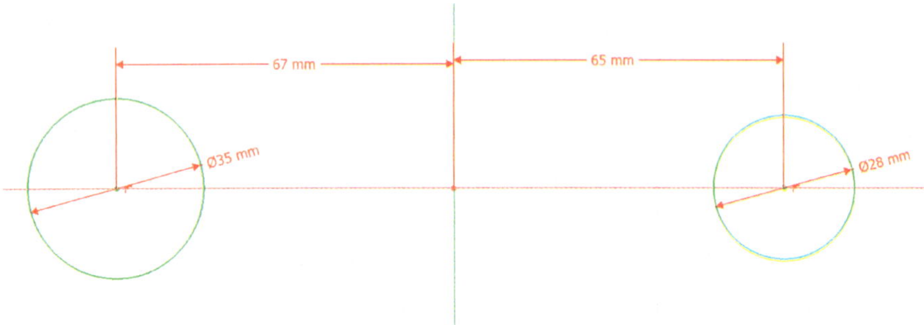

Now we create the middle area. To do this, we first draw two 85 mm long lines, each with a distance of 7.5 mm (vertical) and 42.5 mm (horizontal) from the coordinate origin.

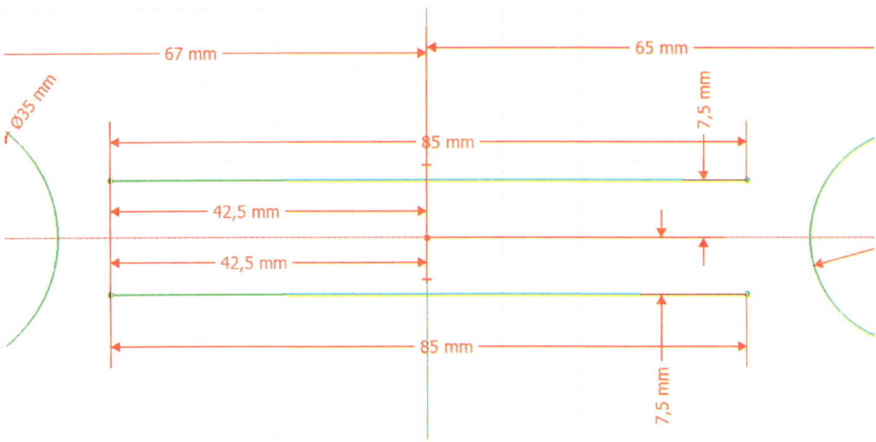

Then we draw two arcs ("End points and rim point") in the left transition area, whose start and end points should lie on the circle and the end points of the horizontal lines, respectively. Similar to the carabiner, we replace the existing constraints at the corner points with tangent constraints ("Constrain tangent").

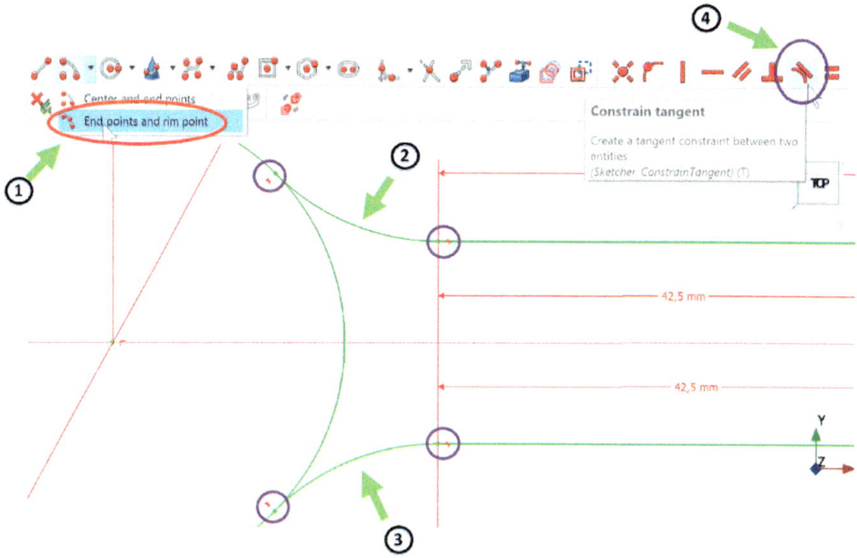

We do the same in the transition area of the right side.

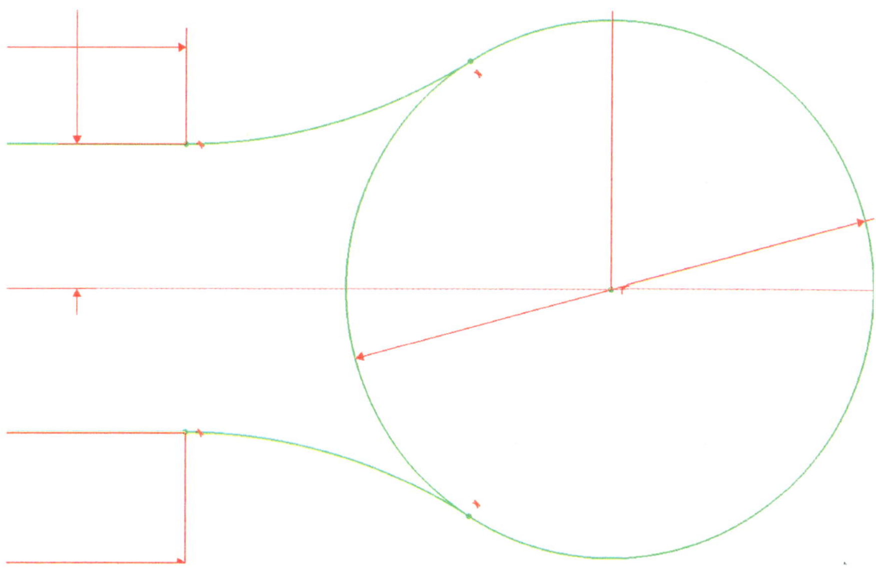

Then we can remove the excess arc segments of the two circles using the tool "Trim edge".

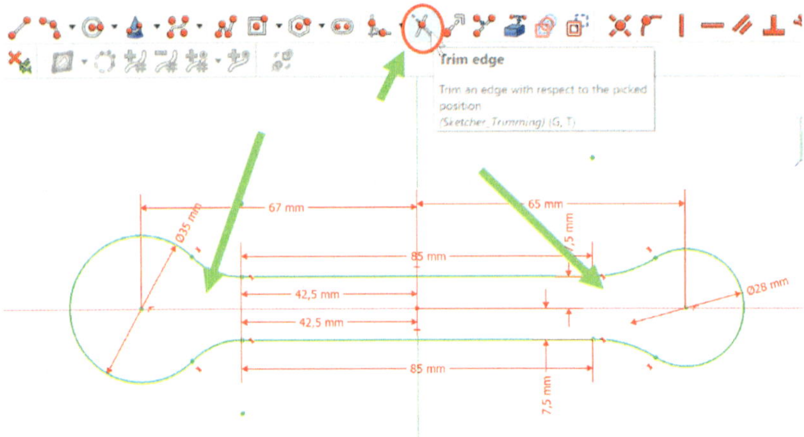

We now turn our attention to the two cutouts that enable the actual function of the wrench. We would like to integrate them into the sketch in order to save one or more work steps. Let's start again in the left area. This geometry is also easiest to sketch with the help of a circle, which we place in the center and do not dimension yet.

We then add a line that should start at the outer circle and be tangential to the inner circle just drawn. Make absolutely sure that the tangential relationship, recognizable by the small symbol, is created. Otherwise, just add it manually. We also need such a line in the lower area. We then set the two lines in parallel dependence with a relationship.

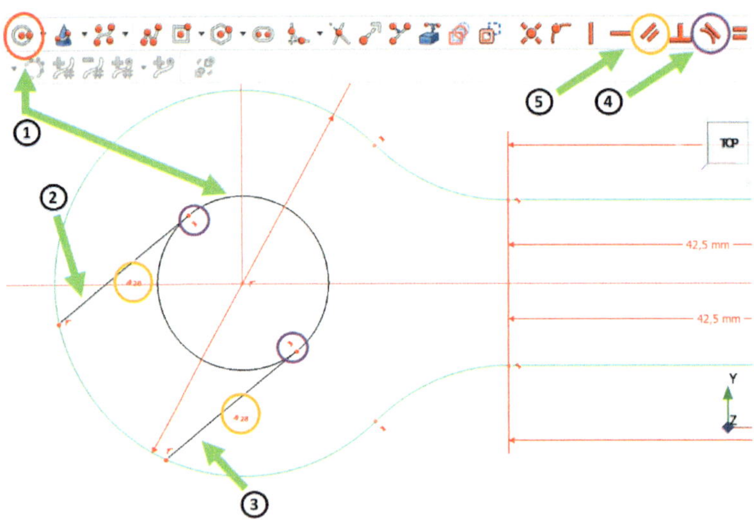

Now we dimension the distance between these two lines with 15 mm. We do this by first selecting the two endpoints of the lines (holding down the CTRL key) and then using the command "Constrain distance". This will give us a 15-wrench size on this side. For a real use, however, the dimensions or tolerances from a table book or the Internet should be used here because there must still be some space between the screw head and the wrench.

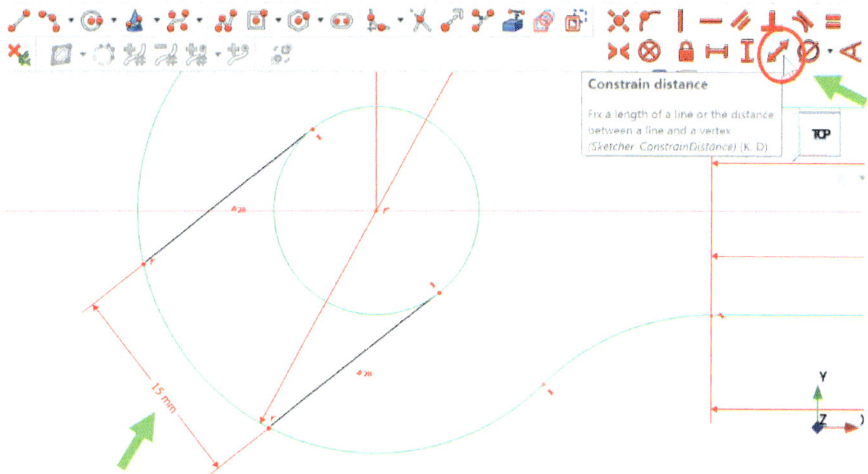

Using the command "Trim edge" we remove the superfluous arc segment in the inner area.

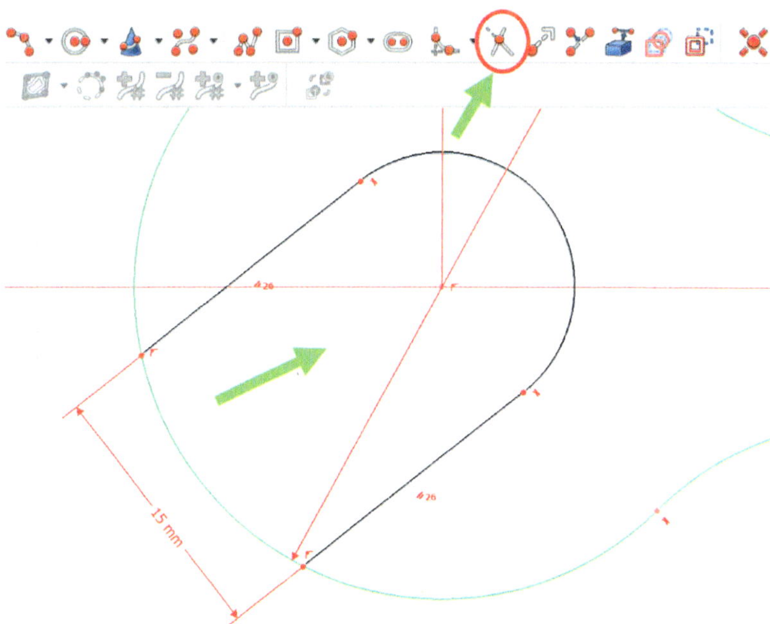

For the complete definition of the sketch, we add an angle of 10° between the upper line and the red horizontal line (x-axis) with the command "Constrain angle". Then the sketch is completely defined again. Finally, we remove the second superfluous arc segment of the outer circle and obtain our opening.

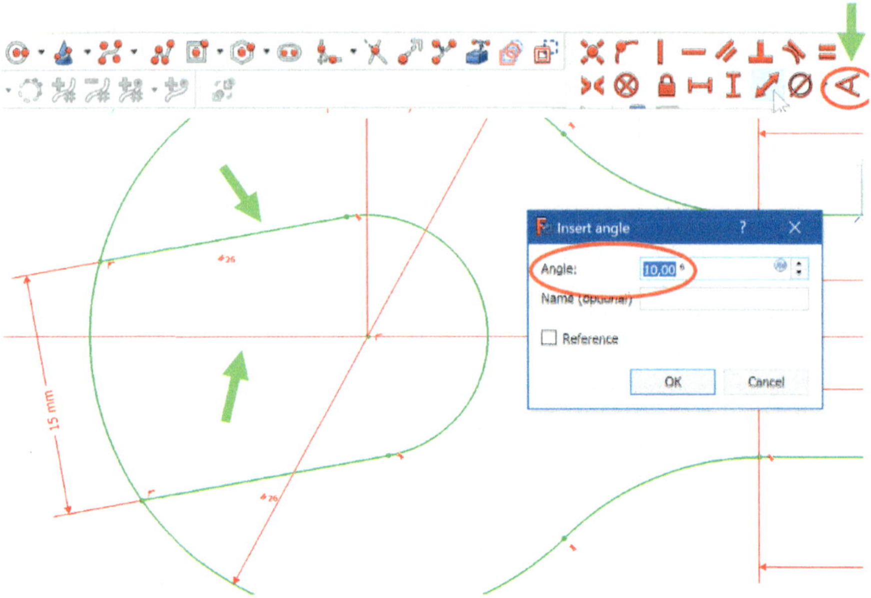

This operation apparently lost a constraint because the sketch becomes black again in the lower area. We add a vertical constraint ("Constrain vertically") between the two points of the tangential arcs.

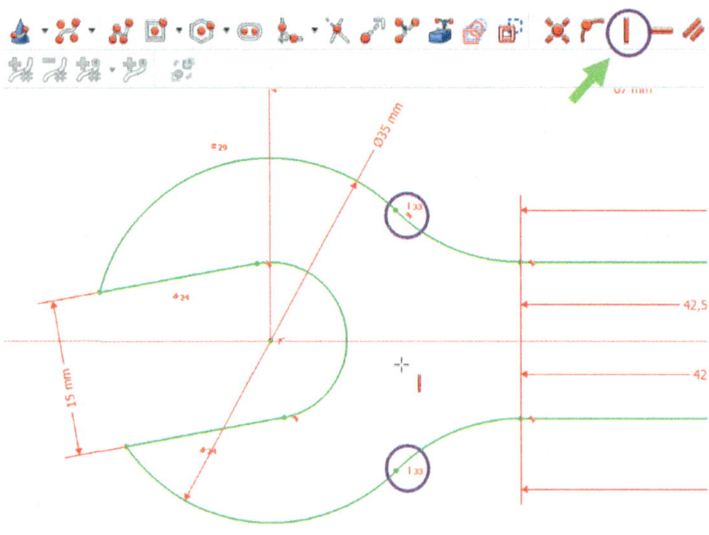

We do the same procedure on the other side, only the dimensions are different, we want a 13-wrench size here. Feel free to try it out on your own! If you get stuck at any point, you can follow the procedure on the previous page.

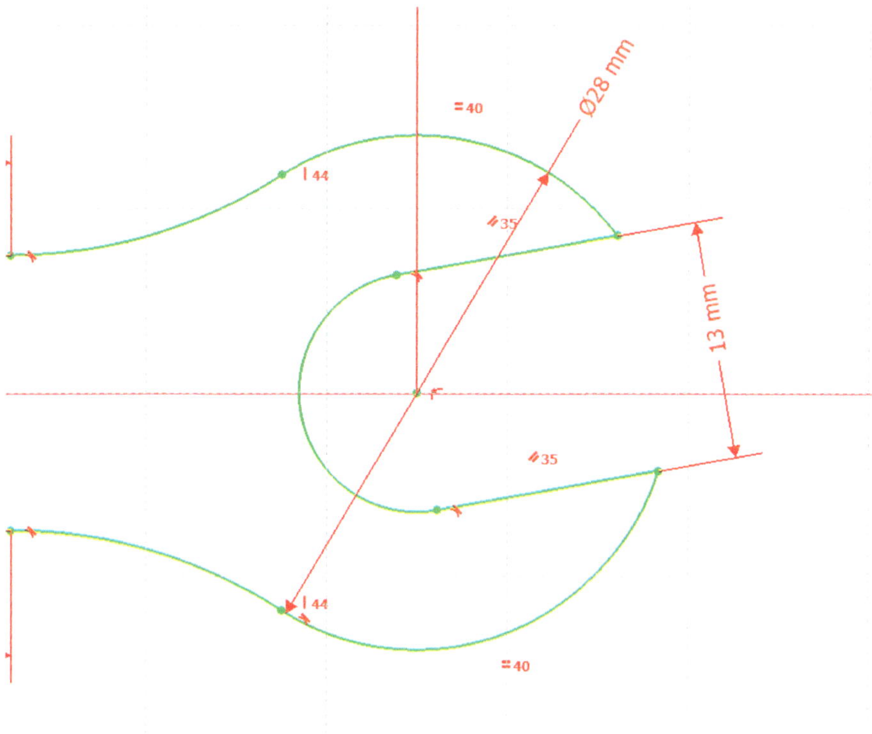

Perfect! The profile is now ready, and we can close the 2D sketch.

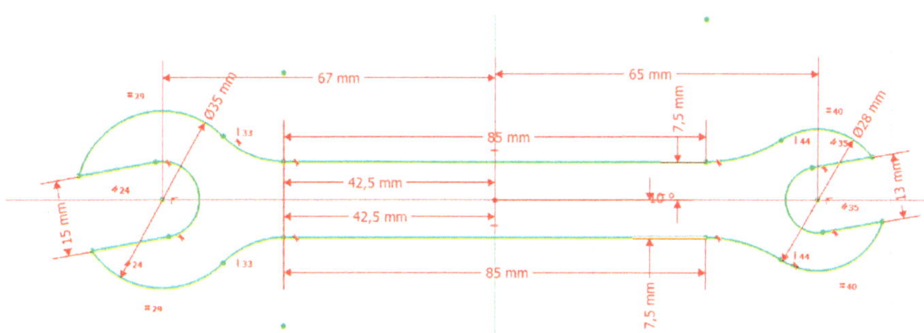

We now simply extrude the profile 6 mm. In this case, however, we extrude 3 mm in each direction, i.e., we select the option "Two dimensions" at the setting "Type". We do this so that the x-y plane of the 3D body is exactly centered within the body. We will see why we need this in a moment.

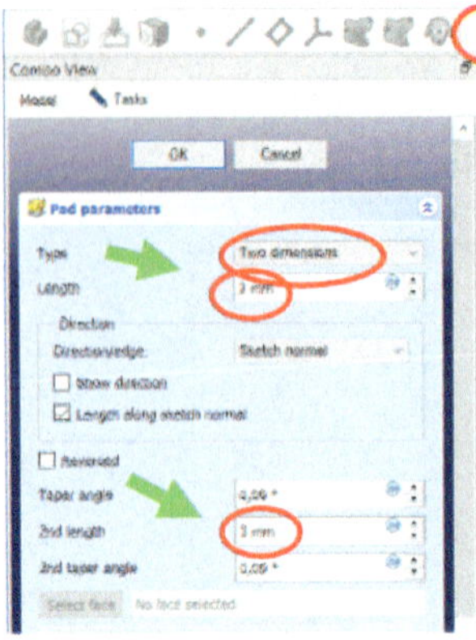

We would now like to add a recess or embossing in the middle area. To do this, we sketch an oblong hole on the upper or lower surface of the 3D model using the "Create a Slot" command. For a better view, we hide the 3D body.

The length of the oblong hole should be 80 mm and the width should be 10 mm. In addition, one of the two points of the oblong hole should lie on the red horizontal line (x-axis) and be dimensioned with 40 mm to the coordinate origin so that the oblong hole is centered.

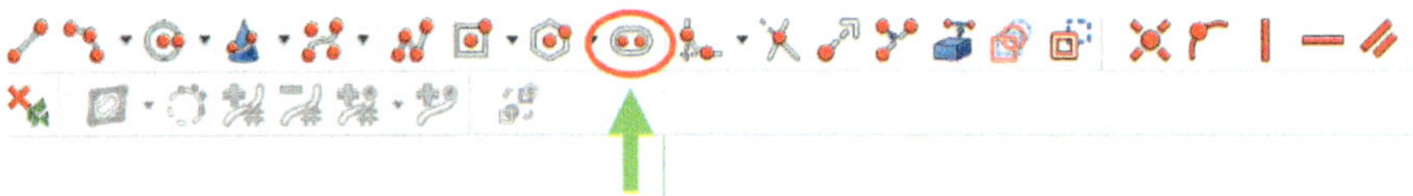

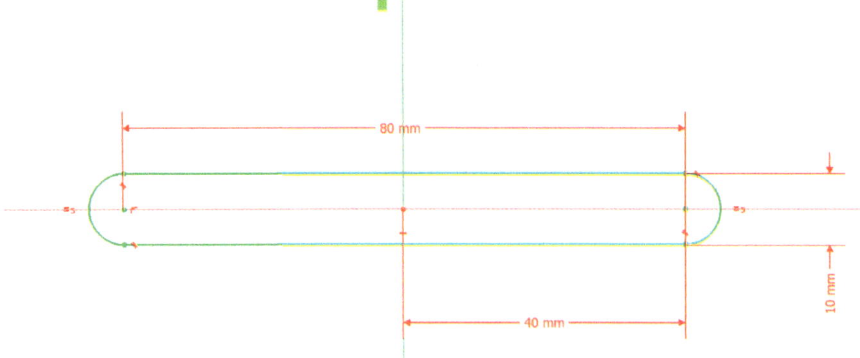

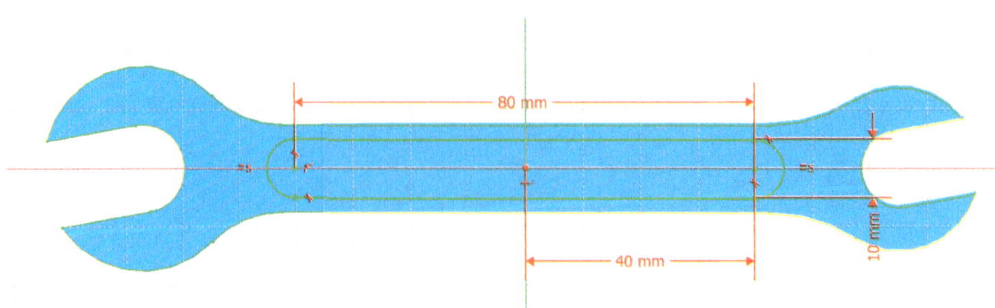

After we close the sketch, we then create a 1 mm deep embossing using the "Pocket" command.

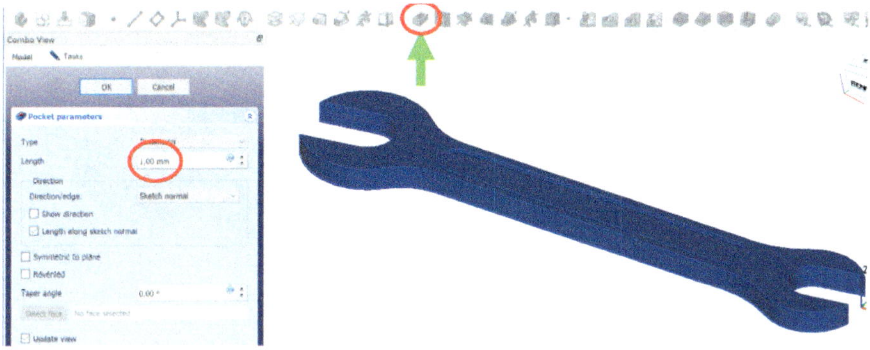

Since we have built the part symmetrically to the x-y plane, we can now easily mirror this indentation to the other side using the "Mirrored" command. We select the recess ("Pocket") in the structure tree and click on the command "Mirrored" in the toolbar. Then in the settings, we select for the mirror plane ("Plane") the x-y plane ("Base XY plane"). This is the same as the extrusion we made earlier in two directions.

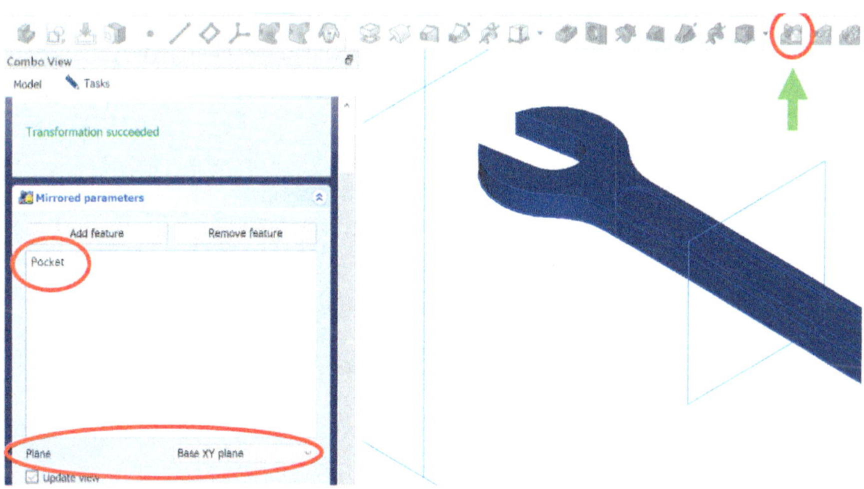

We then round off the four edges of the wrench's receptacles with 2 mm. You can select multiple edges by holding down the CTRL key while selecting them.

Finally, we round all the edges of the upper and lower deck surfaces with a 1 mm radius. It is best to do this in two separate steps.

Of course, we can still change the appearance. For example, we could choose the material steel. The tool will then get a metallic color.

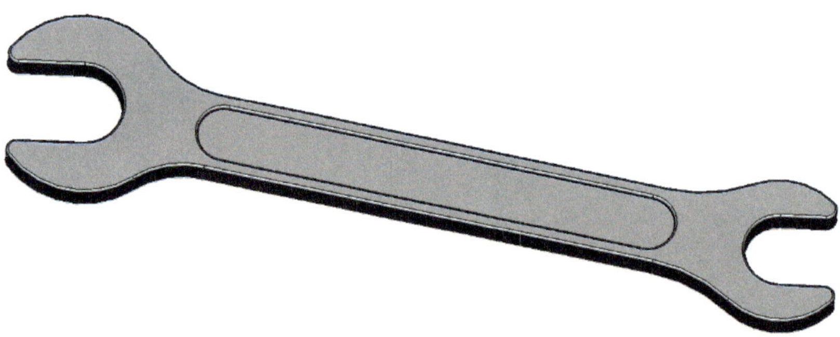

Excellent! The wrench would now be ready for use. Hopefully, you have enjoyed it so far. But of course, this is not the end of the story. Rather, somewhat more complex design projects now follow. On it goes!

8 Project No. 7: Ball bearing

Welcome back! The next design project will be a ball bearing. More specifically, a single row deep groove ball bearing, which is one of the best known and most commonly used ball bearings.

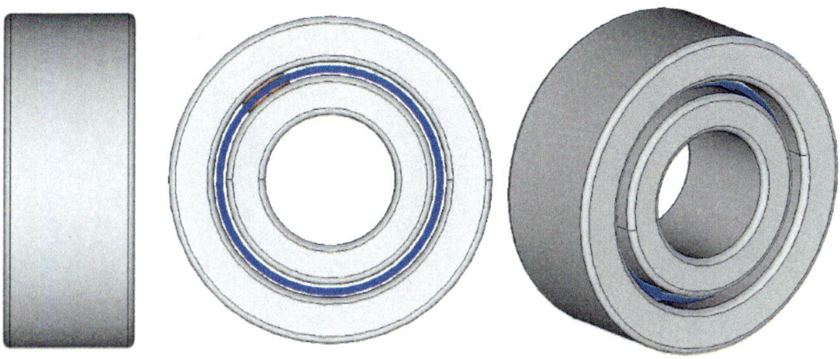

The ball bearing consists of four components. We will create these one after the other. We need an outer ring, an inner ring as well as balls and a ball cage. The so-called ball cage ensures that the balls remain in the correct position.

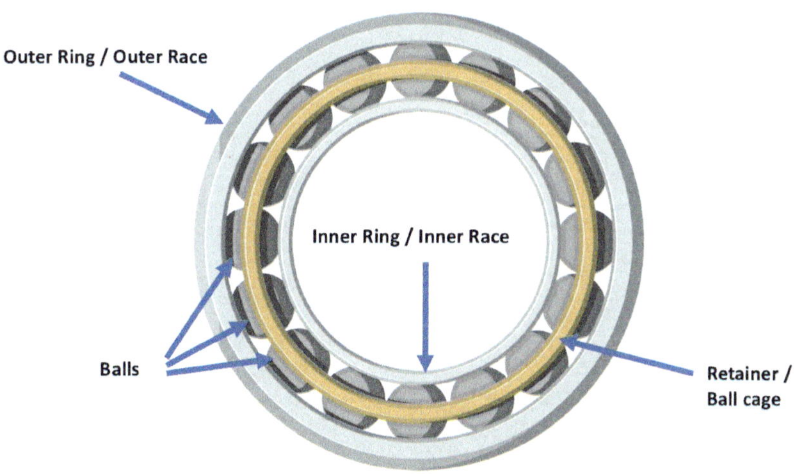

We will start with the first component, the outer ring of the ball bearing, which we will create using a rotation. To do this, we again first need a 2D sketch, e.g., on the x-y plane. For the cross-section of the outer ring, we draw a rectangle 20 mm wide and 7 mm high above the x-axis. It is best to use the command "Centered rectangle" for this so that you can place the center of the rectangle on the green vertical sketch axis. Finally, we add a dimension of 25 mm from the top edge to the coordinate origin.

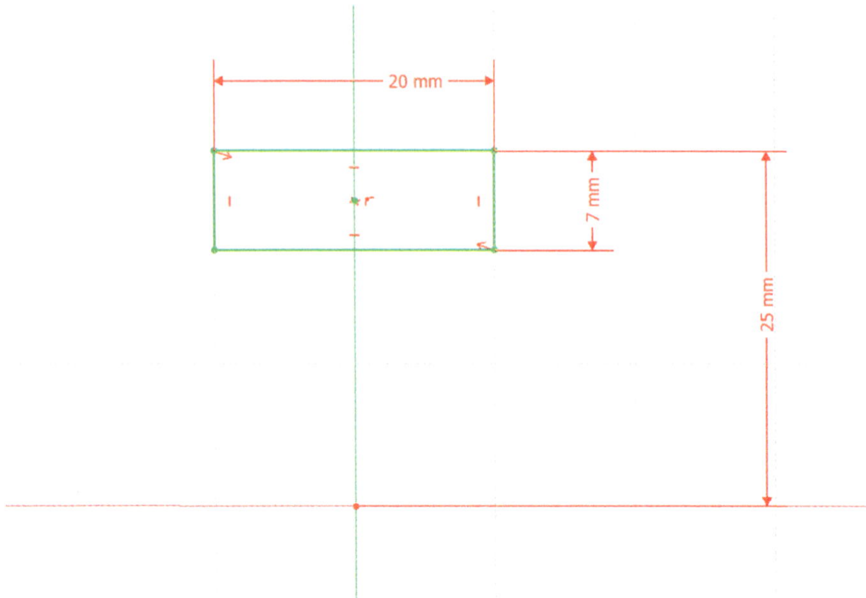

Then we need to create the raceway for the balls. To do this, we use a circle that we place as shown and provide with a diameter of 8 mm. We dimension the distance between the center of the circle and the top edge of the rectangle as 7.8 mm.

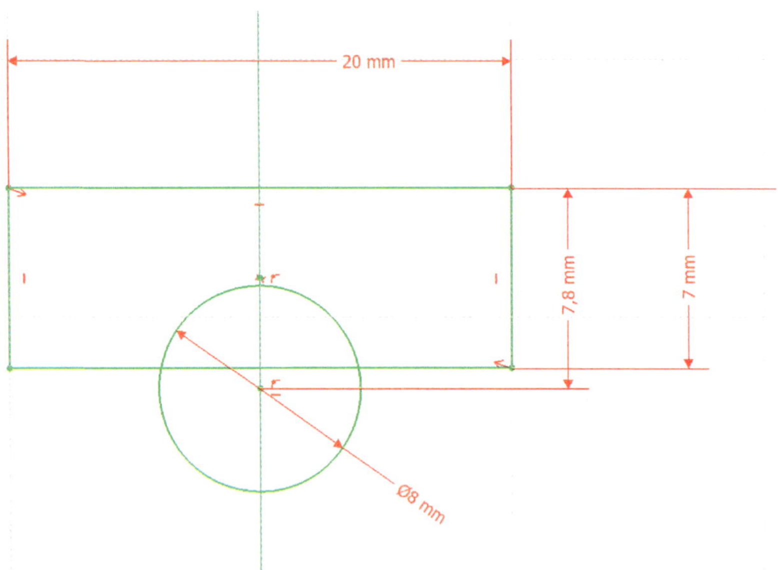

Then we remove two superfluous profile sections with the command "Trim edge". If the complete definition of the sketch is lost as a result, you can fix this by making the two lateral lines of the rectangle identical with the command "Constrain equal".

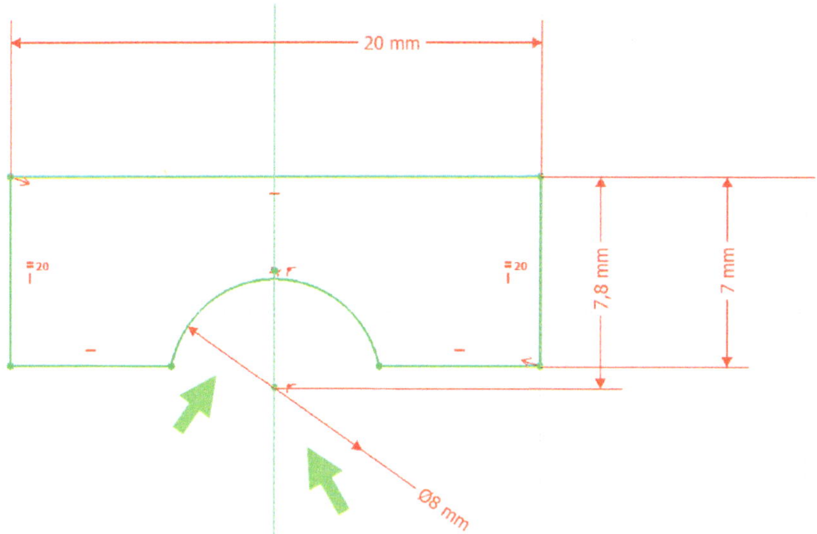

As a last step, we can also create fillets for the edges of the part. To do this, we create fillets with a radius of 1 mm each using the command "Constraint-preserving sketch fillet".

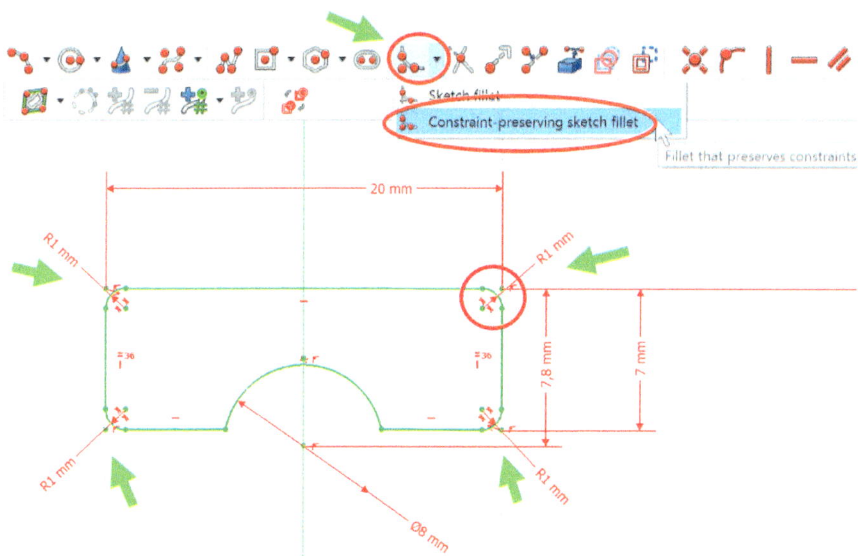

Then the cross-section profile of the outer ring is ready and can be rotated after finishing the sketch with the command "Revolution". For the application of the command to work, we have to select the x-axis as rotation axis in the settings.

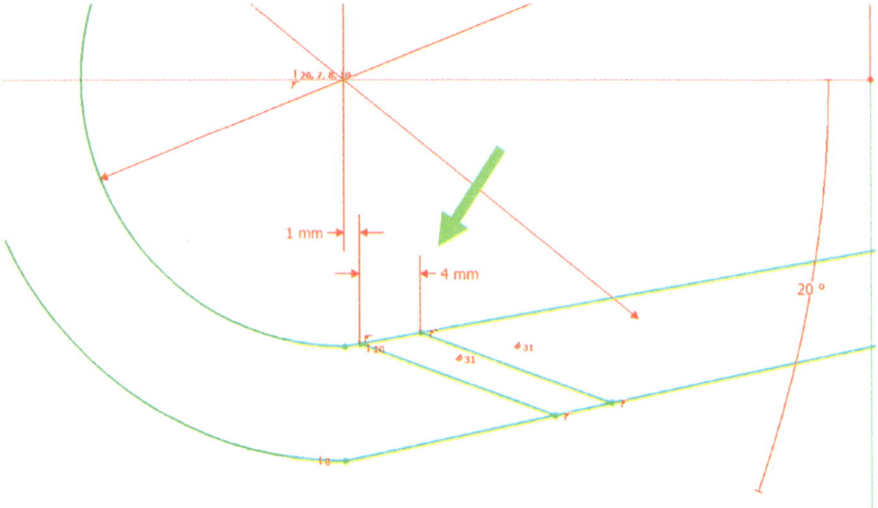

For the second part of the ball bearing, which should be the inner ring, we need to create a new document because it is a separate component. Therefore, we save the outer ring of the ball bearing and close the document. Before that, of course, we can change the appearance of the component, if we wish. We could color it silver, for example.

On the x-y plane of the new document, we will then – analogous to the previous part – also sketch a cross-section geometry, which we will then transform into a 3D component with "Revolution". We start again with a rectangle, which we dimension 20 mm wide and 6 mm high. We define the vertical distance between the origin and the lower edge of the rectangle as 16 mm.

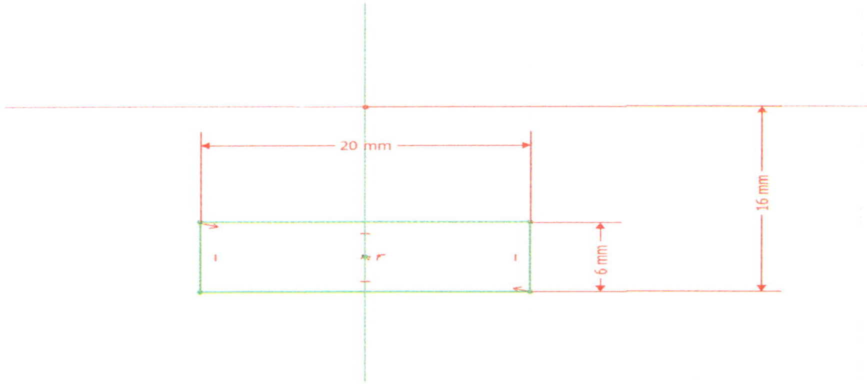

Then we sketch a raceway for the balls here as well. We do this similarly to the outer ring. The diameter of the circle must be identical to the outer ring (8 mm). A distance of 6.8 mm between the center of the circle and the upper edge of the rectangle ensures that the two raceways are concentric to each other.

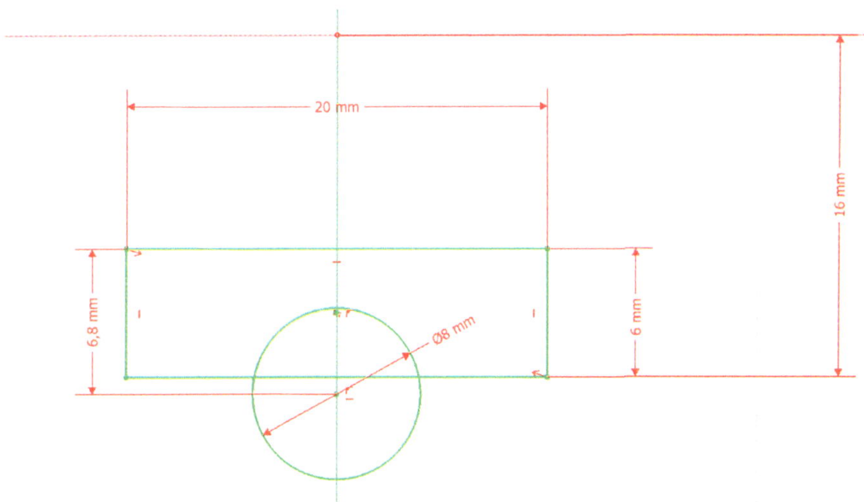

In the last two steps of the 2D sketch, we again remove the superfluous profile sections on the one hand and create 1 mm fillets for the four corners on the other. Also add all necessary constraints for a complete definition of the sketch. The procedure is identical to the outer ring.

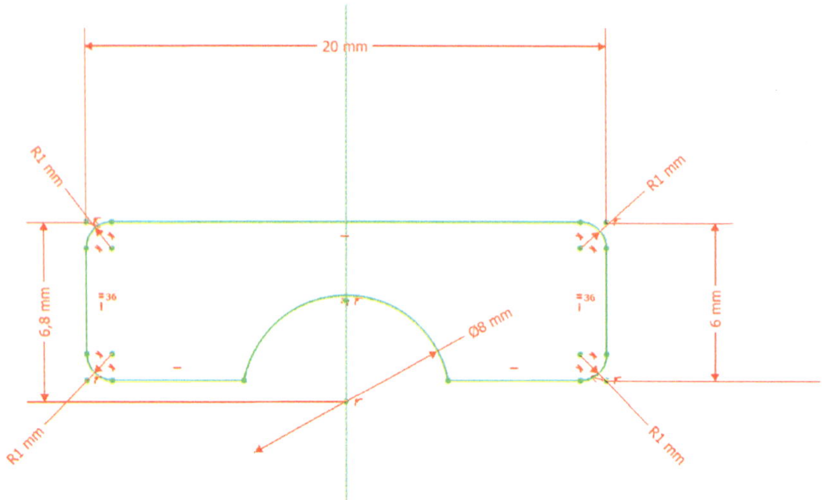

After we close the sketch, we can then make a 360-degree rotation around the x-axis.

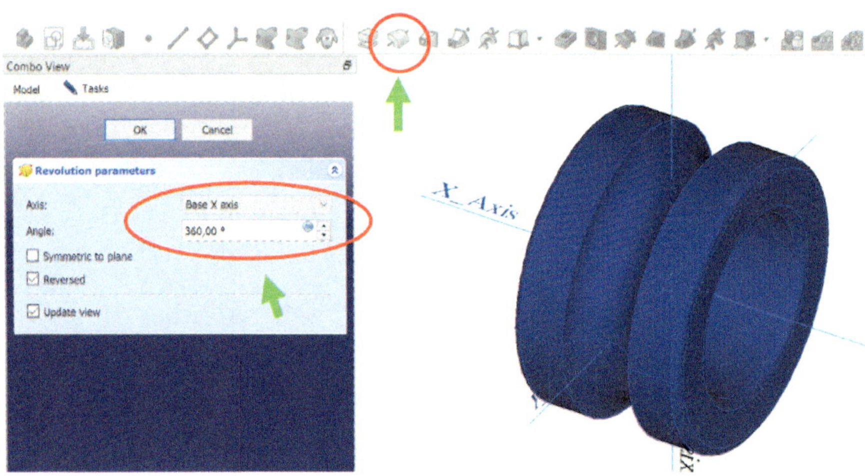

Then – if desired – we can change the appearance of the part, save the part and then close it.

For the next part, the ball cage, we again create a new document, a new body and a sketch, for example on the y-z plane. We do this because this part is also an independent component. However, we will create this part not by a rotation, but by an extrusion. To do this, we simply need to sketch two circles, each of which should start at the coordinate origin and be given a diameter of 33 mm and 35 mm, respectively.

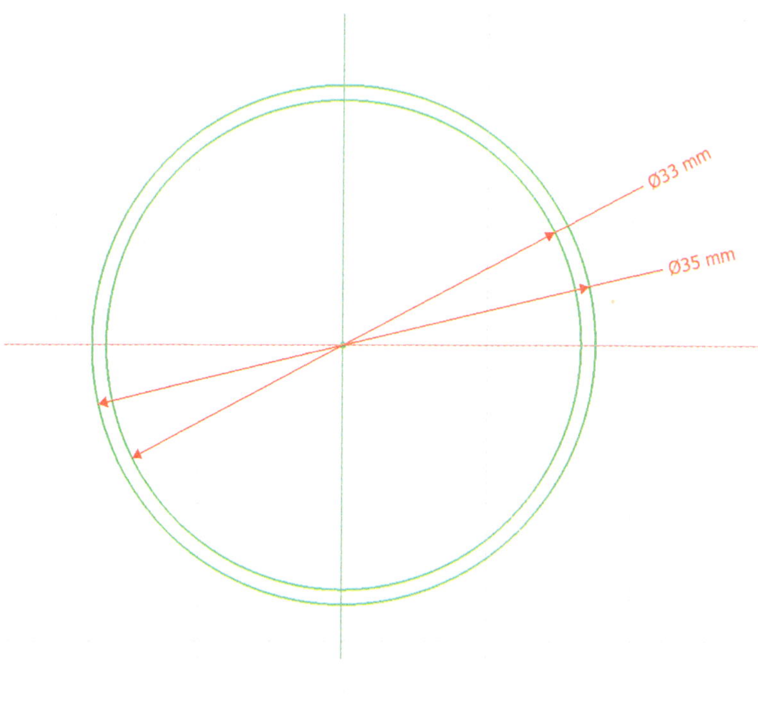

Then we can close the sketch and make a symmetrical extrusion (command "Pad" and setting "Two dimensions") with 6 mm length per direction.

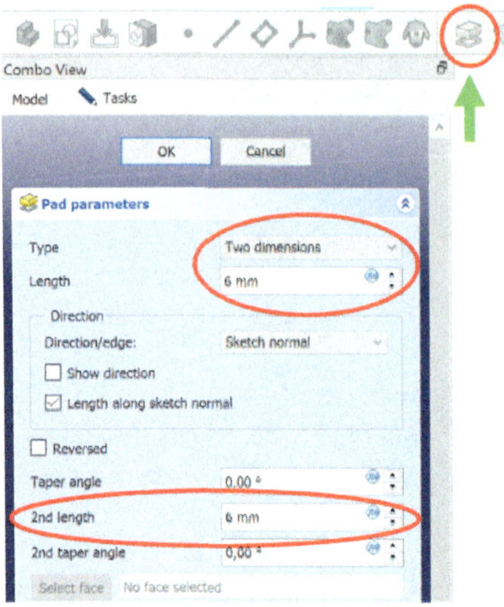

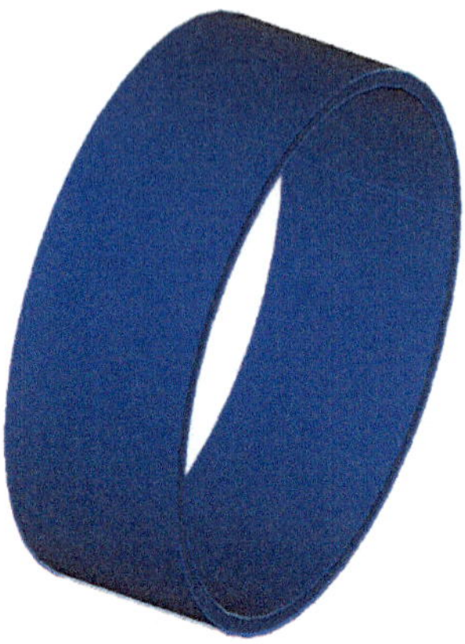

Now we have to add holes where the balls will sit later. To do this, we simply create a circular cutout that should go from the x-y plane up through the body. To do this, we draw a circle with a diameter of 7.8 mm in a sketch on the x-y plane. We can hide the body for this purpose. The center of the circle should be at the center of the coordinates.

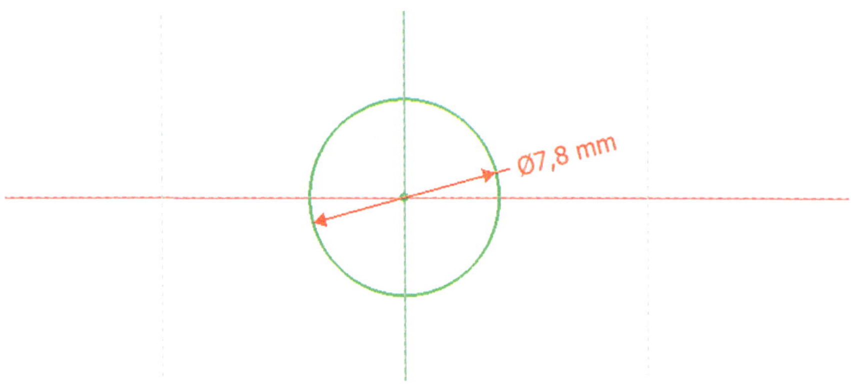

Then we close the sketch and use the command "Pocket" and the setting "Up to face". To do this, we then click on the outer mantle surface of the body. Then the cutout should appear. If the cutout appears in the lower area instead of the upper area, we can change this by activating the option "Reversed".

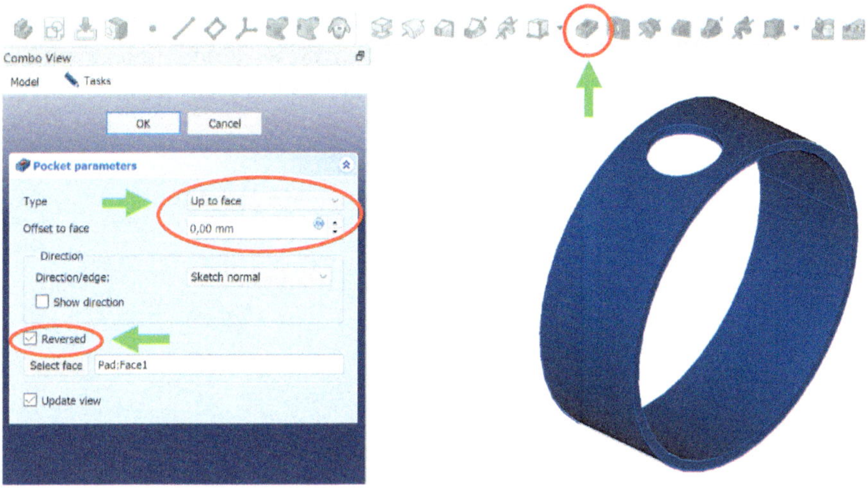

We need nine more holes. To create these easily and quickly, we use the already known function "Polar Pattern". To do this, we first click in the structure tree on the section we have just created ("Pocket") and then on the command "Polar Pattern". The axis for the pattern creation should be the x-axis ("Base X axis") and the number ("Occurrences") should be 10.

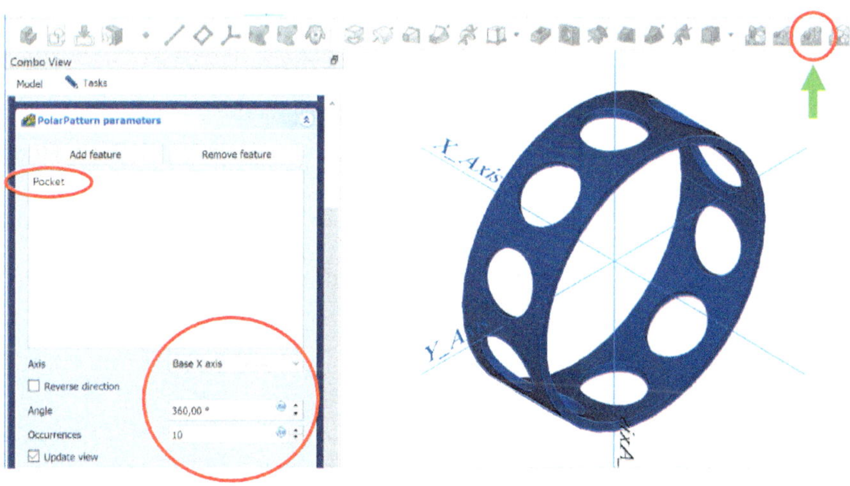

Then the ball cage is ready. We can leave the color as it is or change it. We save the part and close the document.

Now, before we can link all the components together in an assembly in the following, we want to create the last component, the sphere. We need this sphere ten times, but we only need to design it once and then simply insert it ten times into the assembly.

For the sphere, we draw a semicircle with 8 mm diameter in a new document and a sketch on the x-y plane. The center of the semicircle should be on the coordinate origin. In addition, we need a connecting line in the lower area.

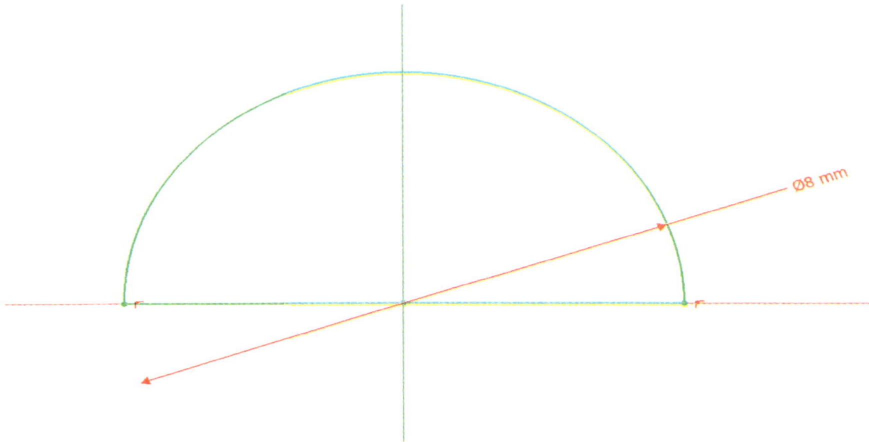

We can then rotate this sketch around the x-axis to create a sphere.

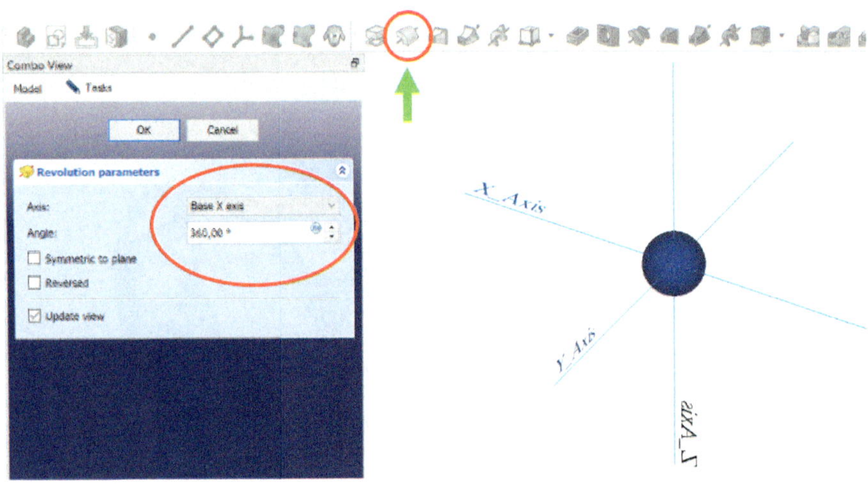

After changing the appearance of the sphere to our liking, we save this document as well.

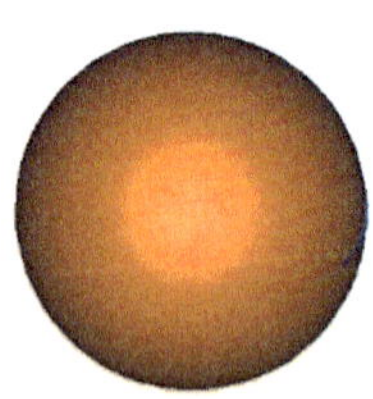

Now we have created all the individual parts of the ball bearing, and we can start assembling these individual parts. To assemble the ball bearing, we create a new document and switch to the "A2plus" workspace. In this workspace, we add – as usual – the first component, which in this case will be the inner ring of the ball bearing. We use the command "Add a part from an external file".

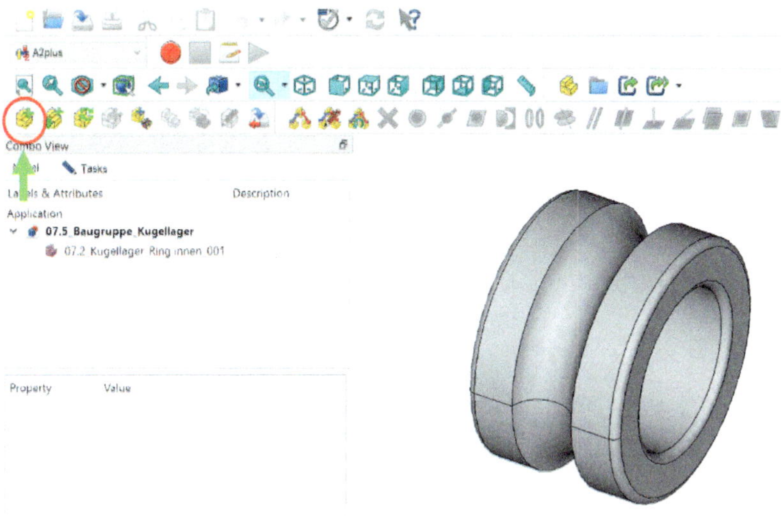

Then we add the sphere cage in the same way and place it anywhere in the workspace. For the first connection of the two parts, we select the two shell surfaces – as shown – and add a concentric constraint with the command "Add axis Coincident constraint".

In order for the ball cage to be centered with respect to the inner ball bearing race, in the next step, we select the two side faces of the individual parts and click on the "Add plane coincident constraint" command.

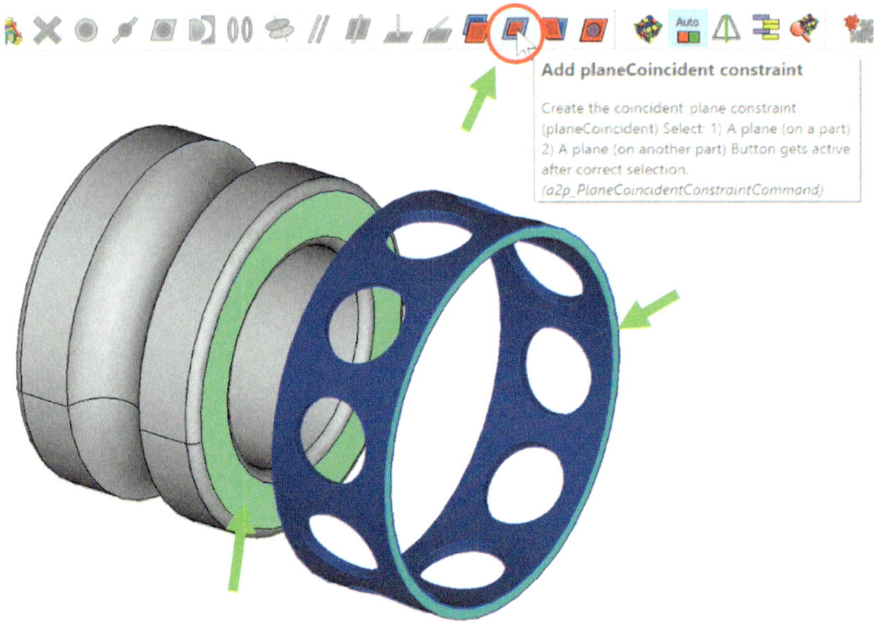

We need an offset of -4 mm so that the ball cage is centered.

Next, we add a sphere to the assembly and link it to the sphere cage by selecting the sphere and the inner lateral surface of a hole and then selecting the command "Add pointOnLine constraint".

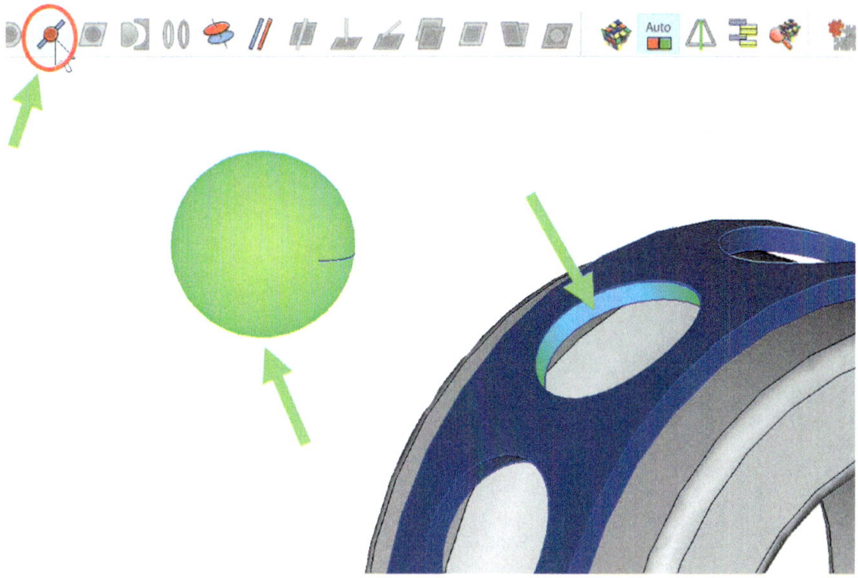

Using the button "Move the selected part under constraints" we can then move the ball to the desired position.

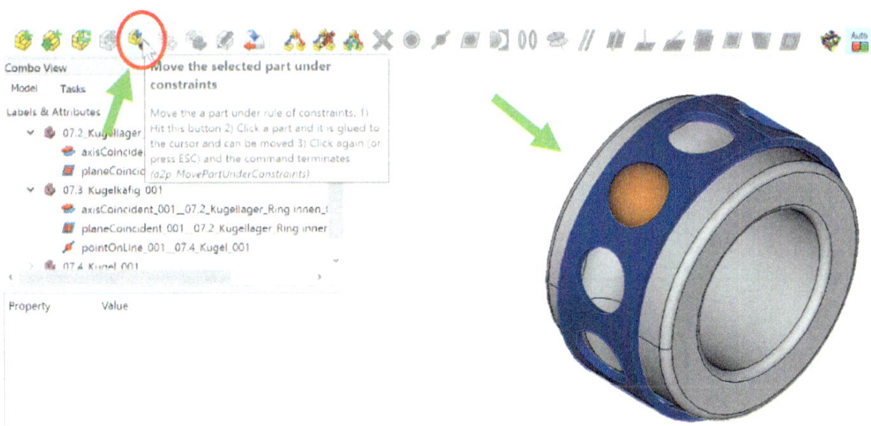

We would now have to repeat this in the same way for all the other nine balls. You are welcome to do this on your own. We'll continue with the outer ring of the ball bearing in a moment. We'll add this to the assembly first.

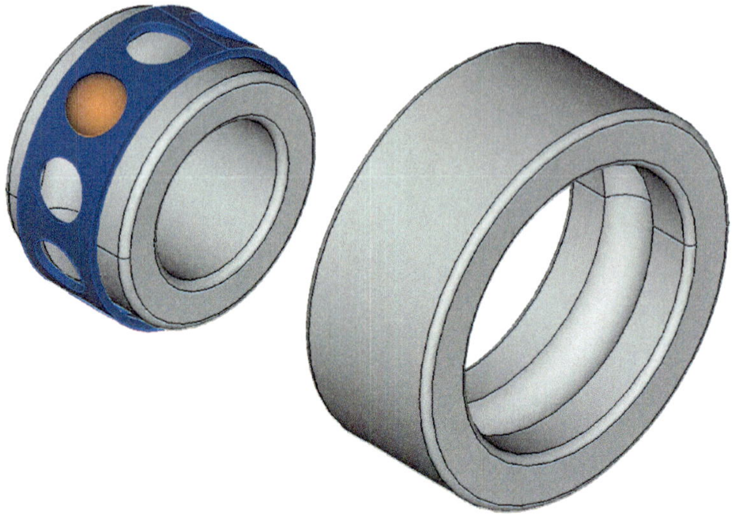

The linkage is now done in the same way as for the linkage between the ball cage and the inner ball bearing race. The only difference here is that we do not need an offset. You are welcome to try this out on your own. By the way, this mating procedure also usually leads to the goal with other assemblies.

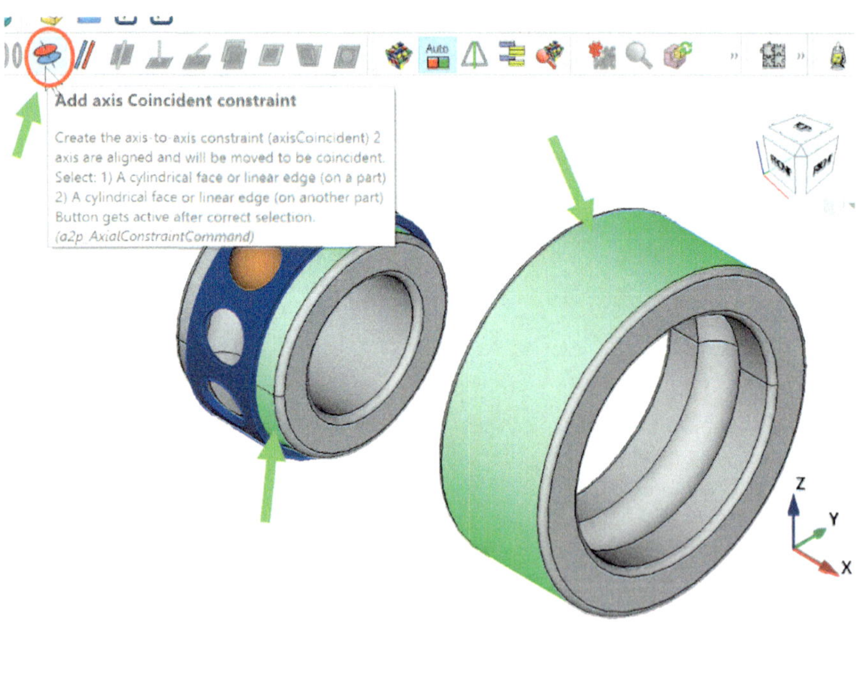

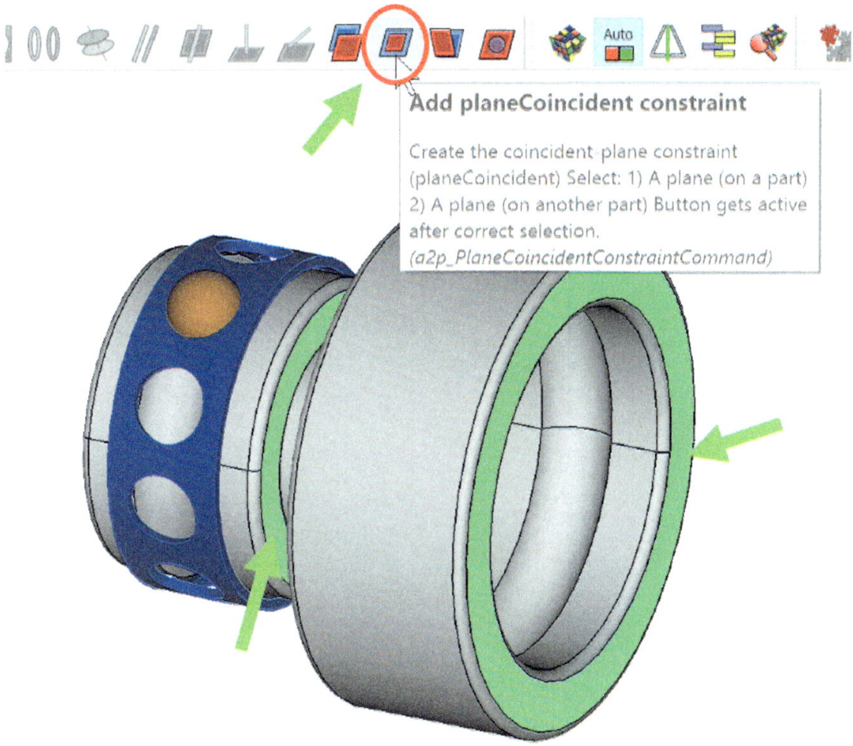

The next design project will be a watering can, before we will then design a remote control. So, we still have a lot to do! Let's move on.

9 Project No. 8: Watering can

Now let's move on to the next design project. In this one, we would like to construct a designer watering can.

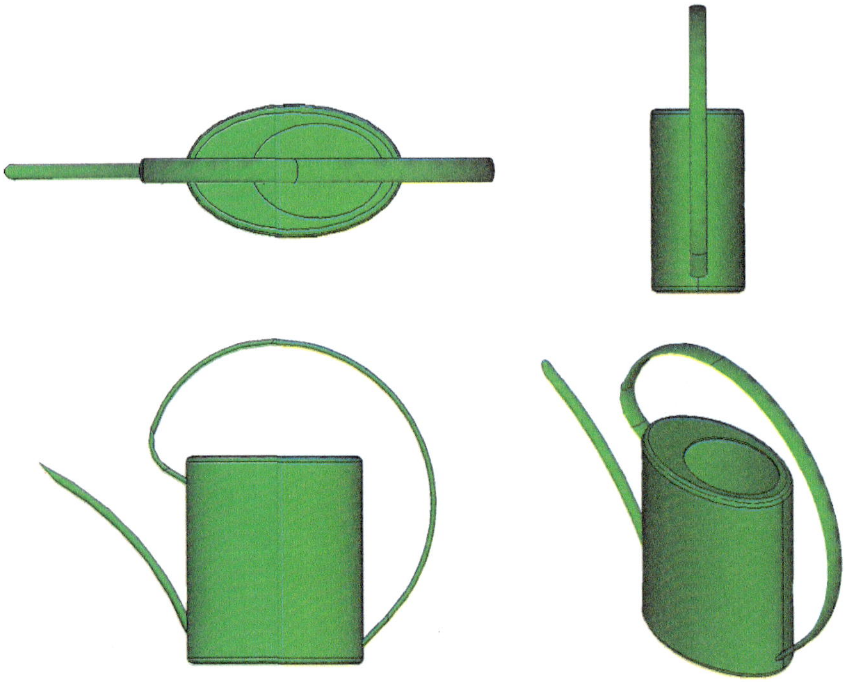

If we mentally break down the finished watering can into its individual parts, we can see that we need an oval and hollow base body with a recess in the upper area, as well as a neck in the front area and a handle. We will then add the latter two elements to the base body later.

It is always very helpful to imagine individual basic bodies and to think about how to build them up. We create a 2D sketch on the x-y plane for the basic oval body we want to create using an extrusion. In the 2D sketch we select the command "Ellipse by center, major radius, point" and then click successively on the coordinate origin, the red horizontal axis (x-axis) and finally on any point in the drawing plane.

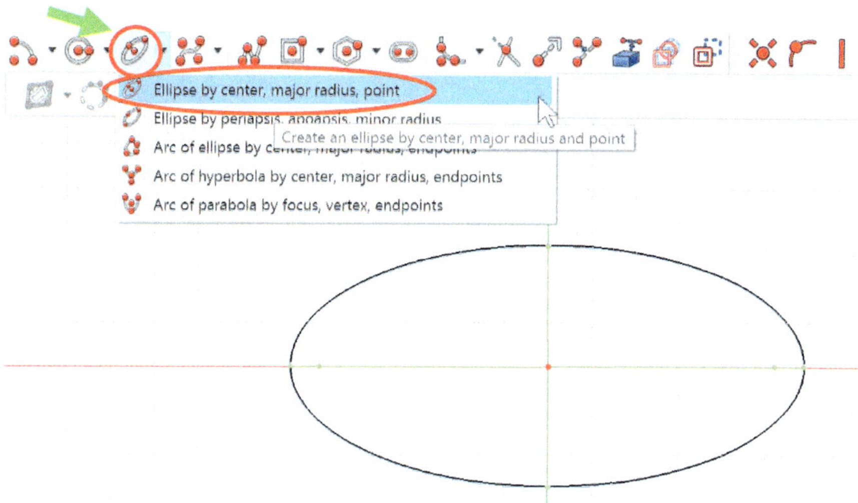

We then dimensioned the width of the ellipse as 140 mm and the height as 85 mm.

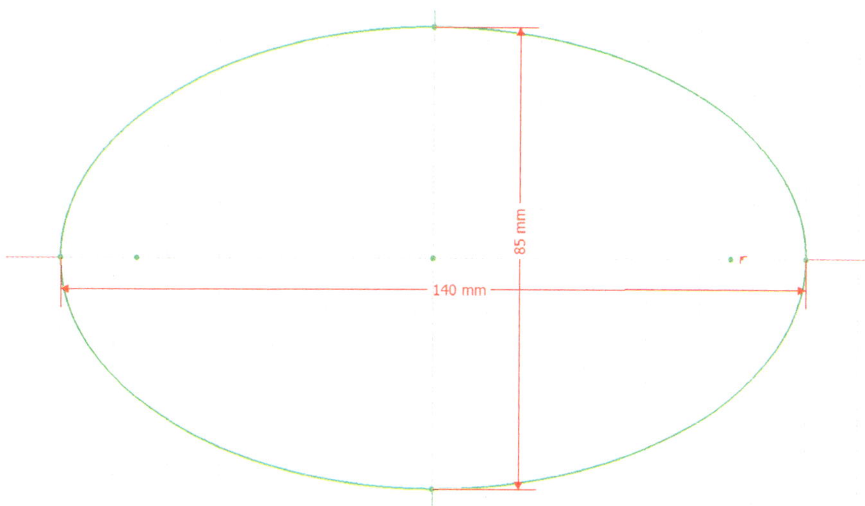

Now we can finish the 2D sketch. We now use the function "Pad" as usual to create the base body. The watering can is to be 160 mm high.

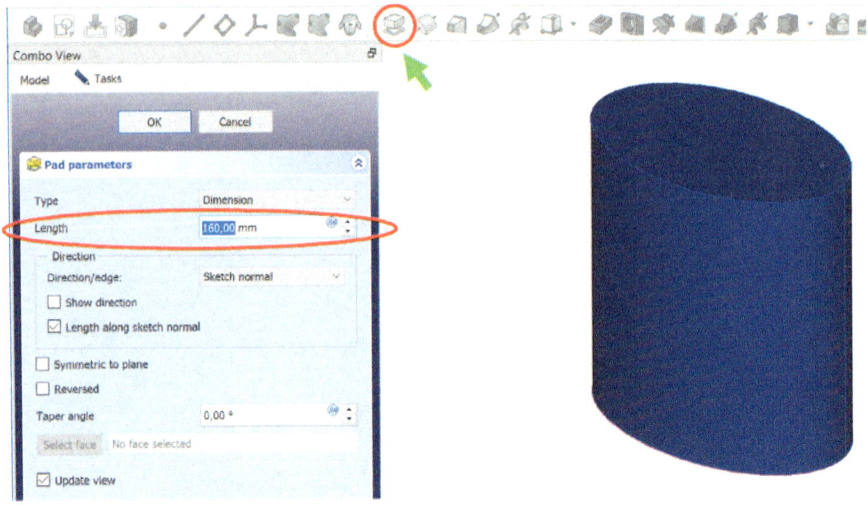

To be able to attach the front neck of the watering can, in the next step, we first create a parallel plane to the y-z plane with 65 mm distance. We do this with the command "Create a datum plane". Please enter a dimension of -65 mm in the option "In z-direction". We use this dimension because the neck must start slightly inside the watering can, to ensure a correct transition, as we will see later.

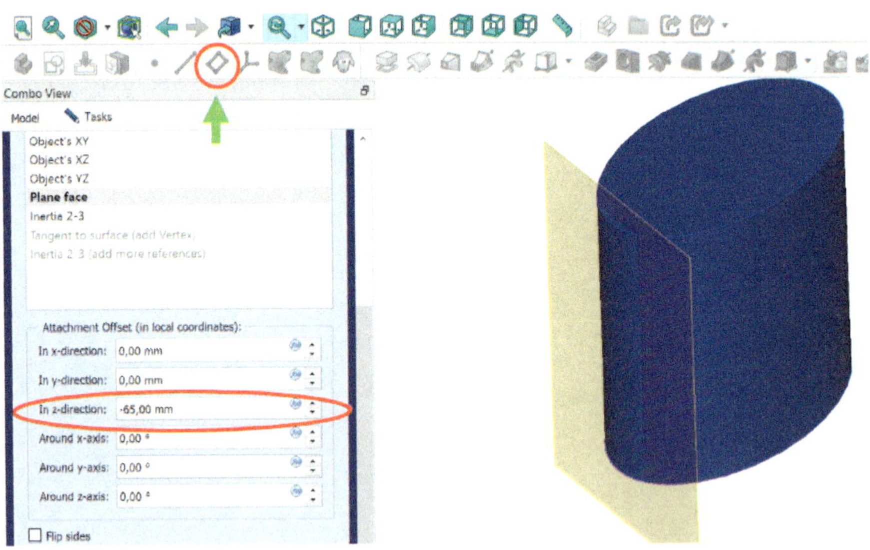

On this plane, we then sketch an ellipse again as the basic profile for the watering can neck. The center of this ellipse should sit on the vertical sketching axis and 20 mm above the bottom of the watering can. The dimensions of the ellipse should be as follows: 10 mm wide and 20 mm high. It is best to hide the base body before drawing.

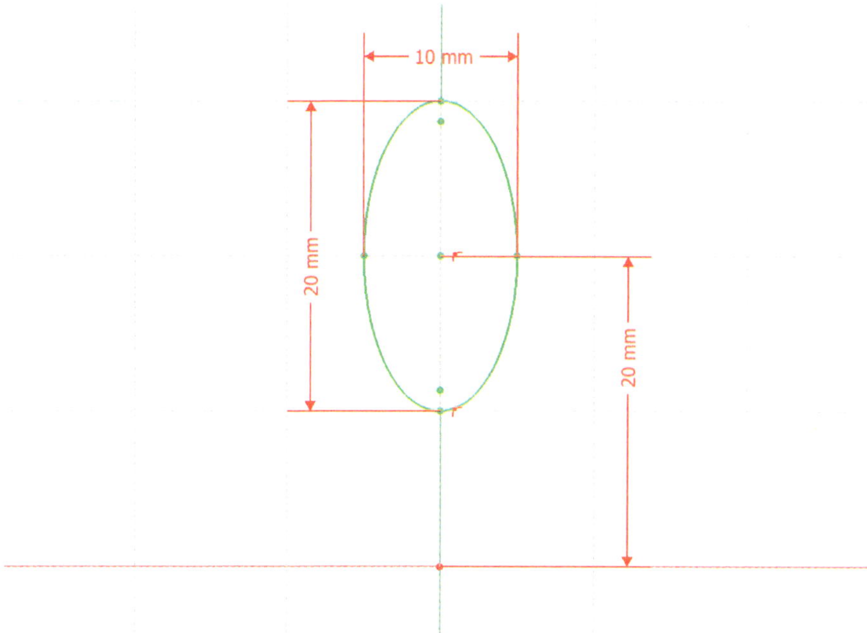

Then we can finish the sketch. We want to create the watering can neck using the function "Additive Pipe".

As you may remember from the beginner's course, we always need a profile and a path for this function. Before we draw this path, we add the front boundary of the watering can neck.

To do this, we create another plane that should be parallel to the y-z plane with a distance of -180 mm.

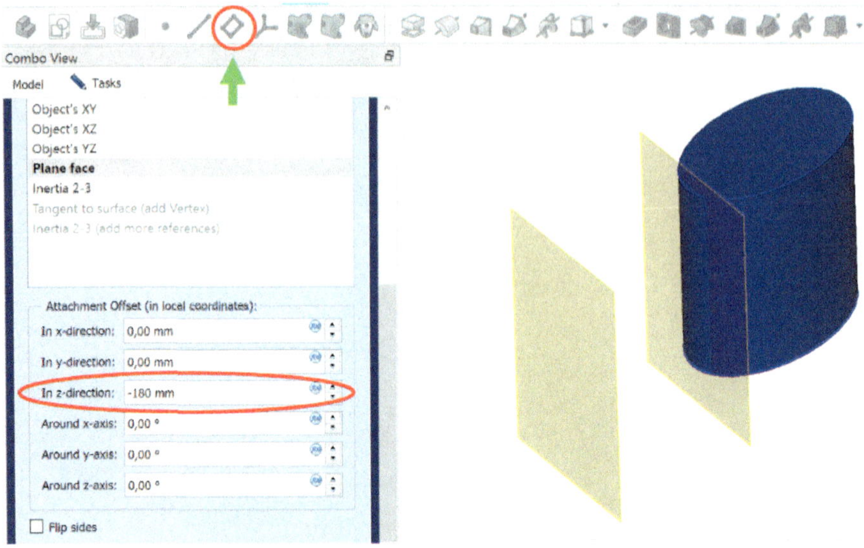

On this plane, we draw another ellipse, which should be 8 mm wide as well as 10 mm high. In addition, the center of the ellipse must lie on the green sketching axis and have a distance of 150 mm from the coordinate origin.

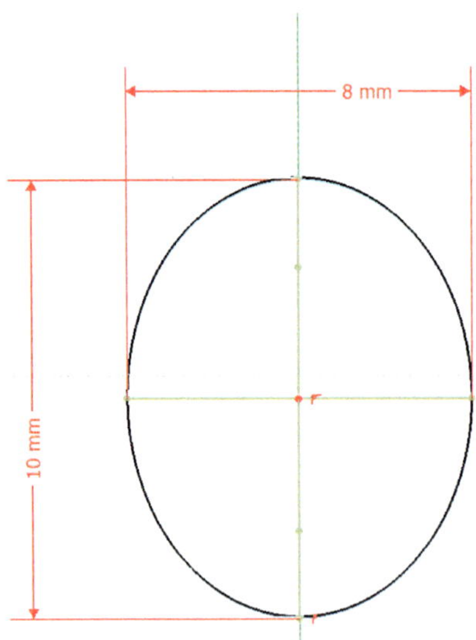

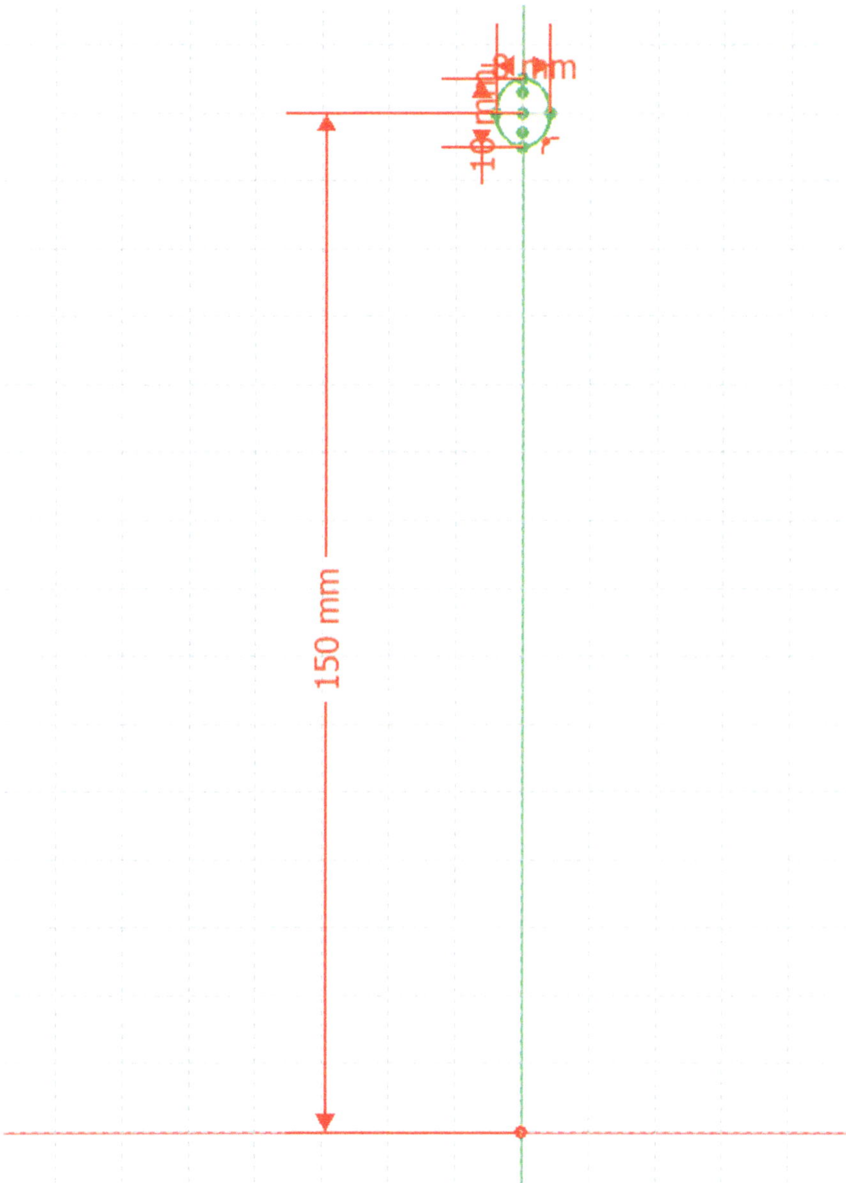

After finishing this sketch, we can start a new sketch on the x-z plane, in which we draw the path for the "Additive Pipe" command. The path consists of an arc segment ("End points and rim point") to which we add the following dimensions. The dimensions each have the coordinate origin as reference point (except for the radius).

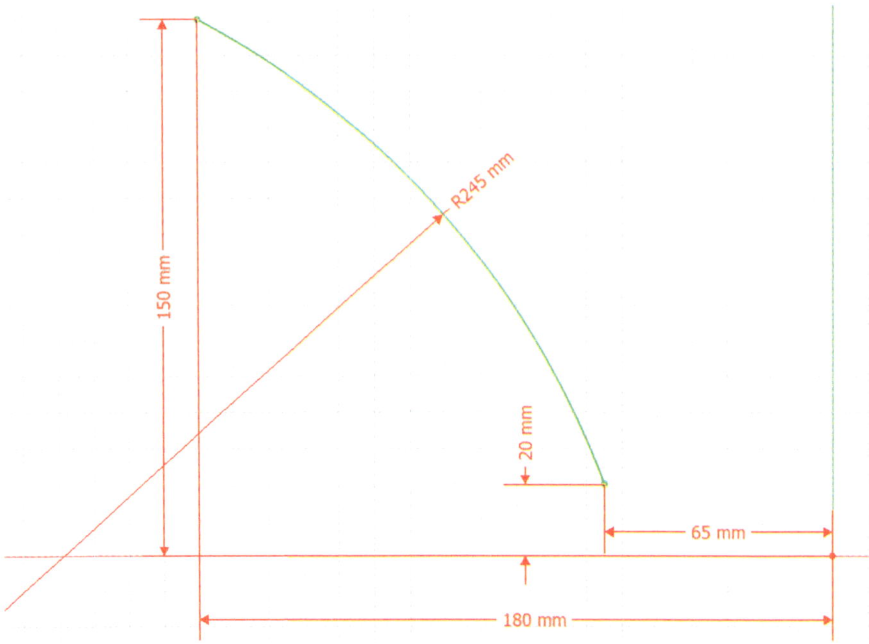

Then we can also finish this sketch. To execute the command, we first select the two sketches with the ellipses in the structure tree and then click on the "Additive Pipe" command in the toolbar.

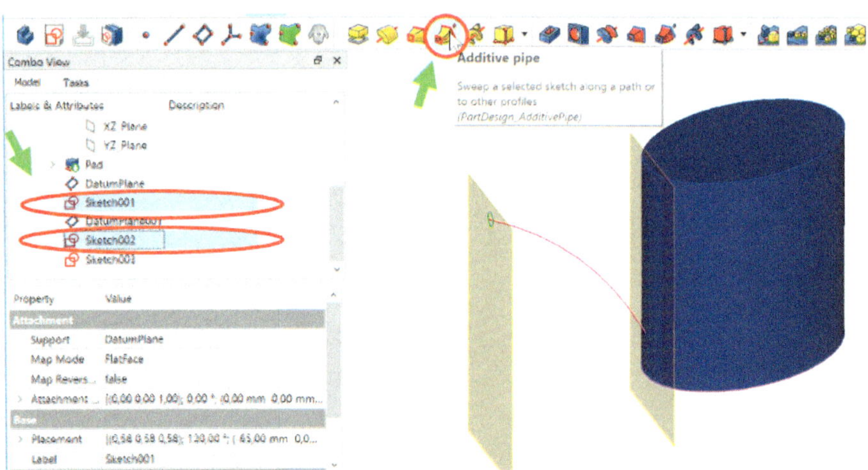

In the next step, we have to determine the path for the command by clicking on the button "Object" in the settings in the field "Path to sweep along" and then selecting the sketched arc in the drawing plane. The program should then create the body as desired.

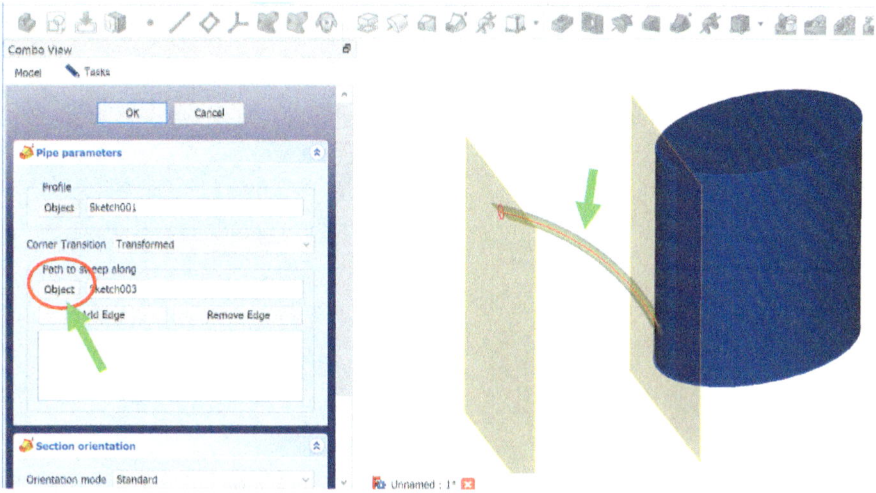

After that, we want to hollow out the base body. How do we do that? Exactly, with the "Thickness" command! To do this, first select the top face and the top face of the watering can neck (hold down the CTRL key) and then click on the command in the toolbar.

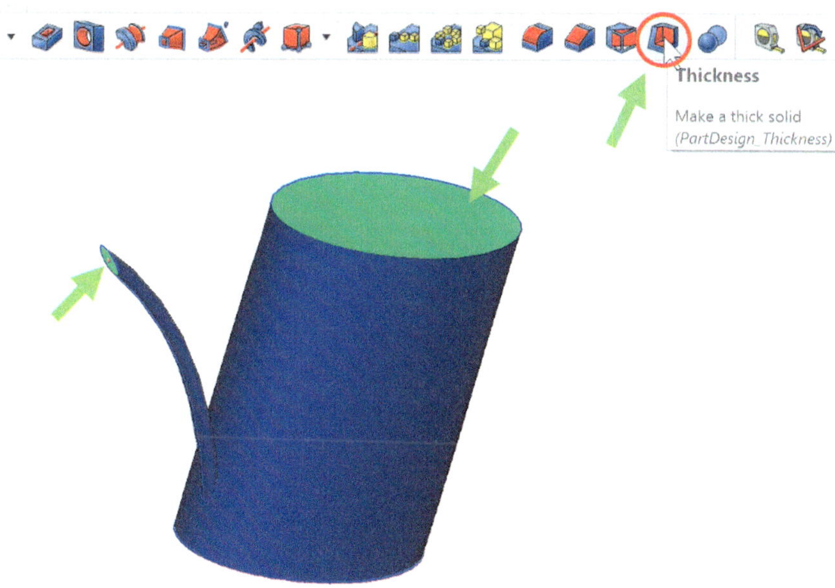

We need a wall thickness of 1.5 mm.

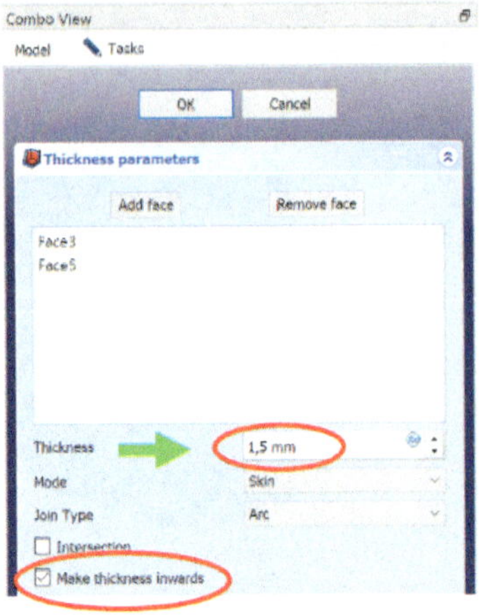

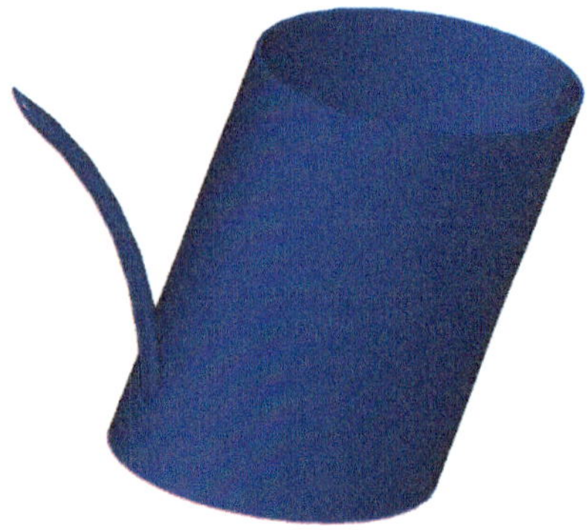

In the next step, we create a lid including a filling opening. To do this, we draw the following profile in a new sketch on the upper edge of the watering can.

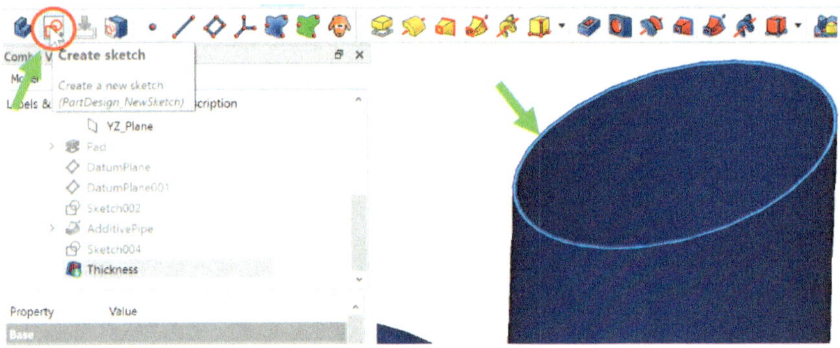

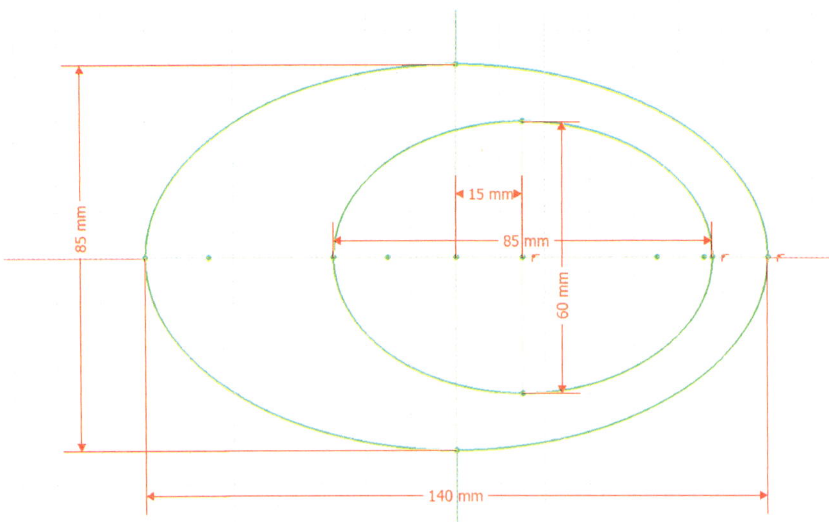

With the help of the profile and the function "Pad" we create a 1.5 mm high extrusion.

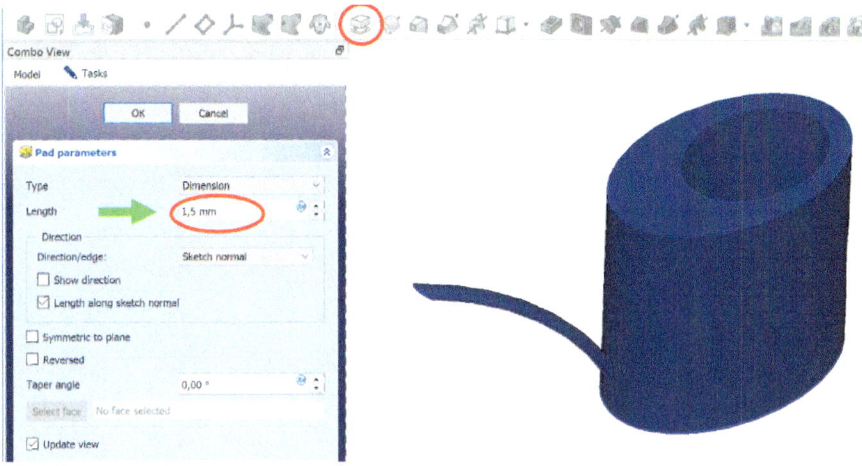

Now we are very far, only the handle is missing. We create the handle in a similar way to the watering can neck. Again, we use the function "Additive Pipe". As profile, we draw an ellipse in the back area on an offset-plane, which should have a distance of 68 mm to the y-z plane.

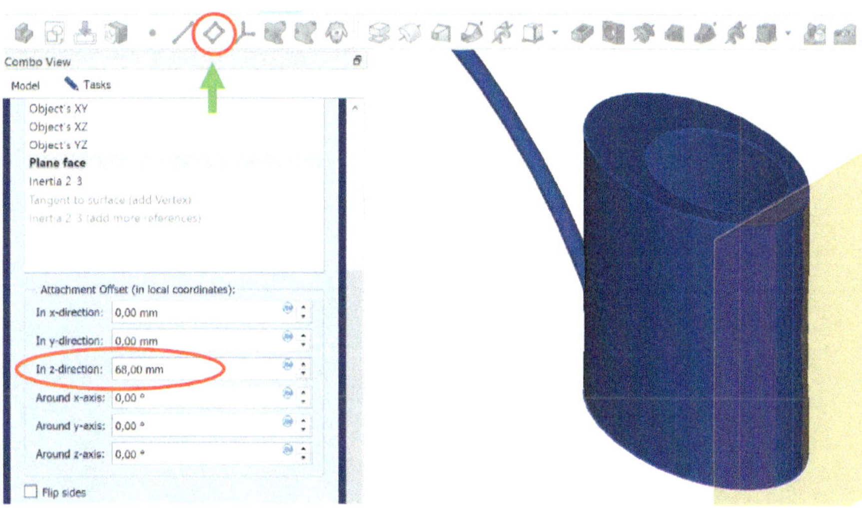

The ellipse we draw on the offset plane should be 15 mm wide and 7 mm high, and have a vertical distance from the origin of 15 mm.

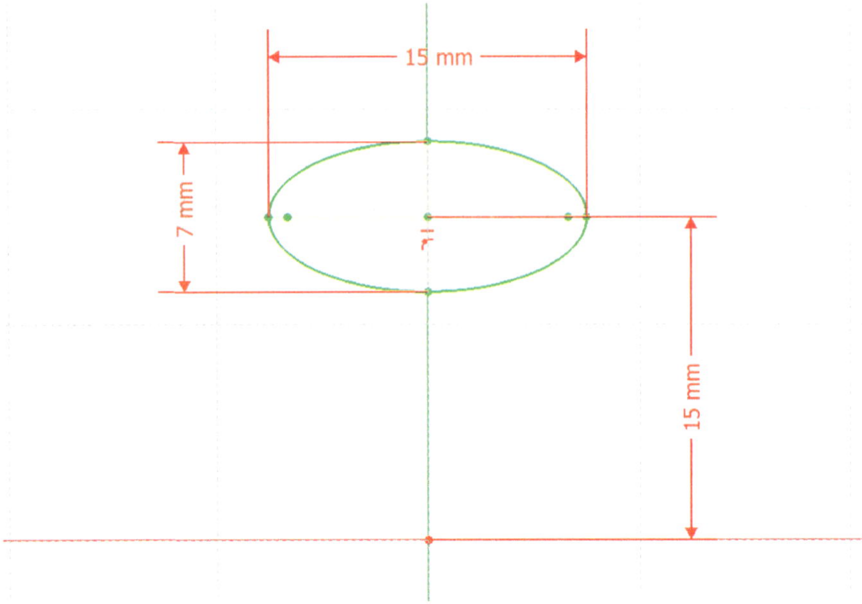

We finish the sketch and then create the path by starting a sketch on the x-z plane. In terms of design, the handle of the watering can should be relatively tapered and curved. To do this, we first draw a simple line that should start 68 mm (horizontally) or 15 mm (vertically) from the origin. This line should have an angle of 135° to the red sketching axis and be 30 mm long.

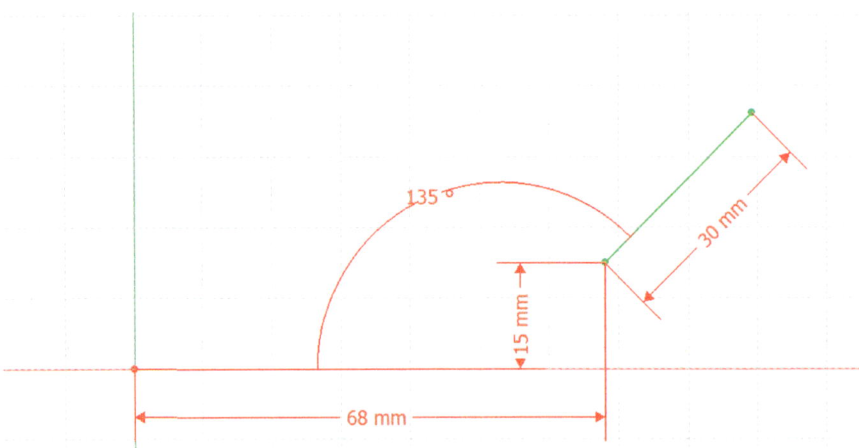

For the further path of the handle, we draw a very curved profile. For this, we add a 3-point arc ("End points and rim point") between the end point of the 30 mm long line and the vertical sketch axis. In the lower area, the arc should sit tangential to the line (create constraint "Constrain tangent") and have a distance of 250 mm to the coordinate origin.

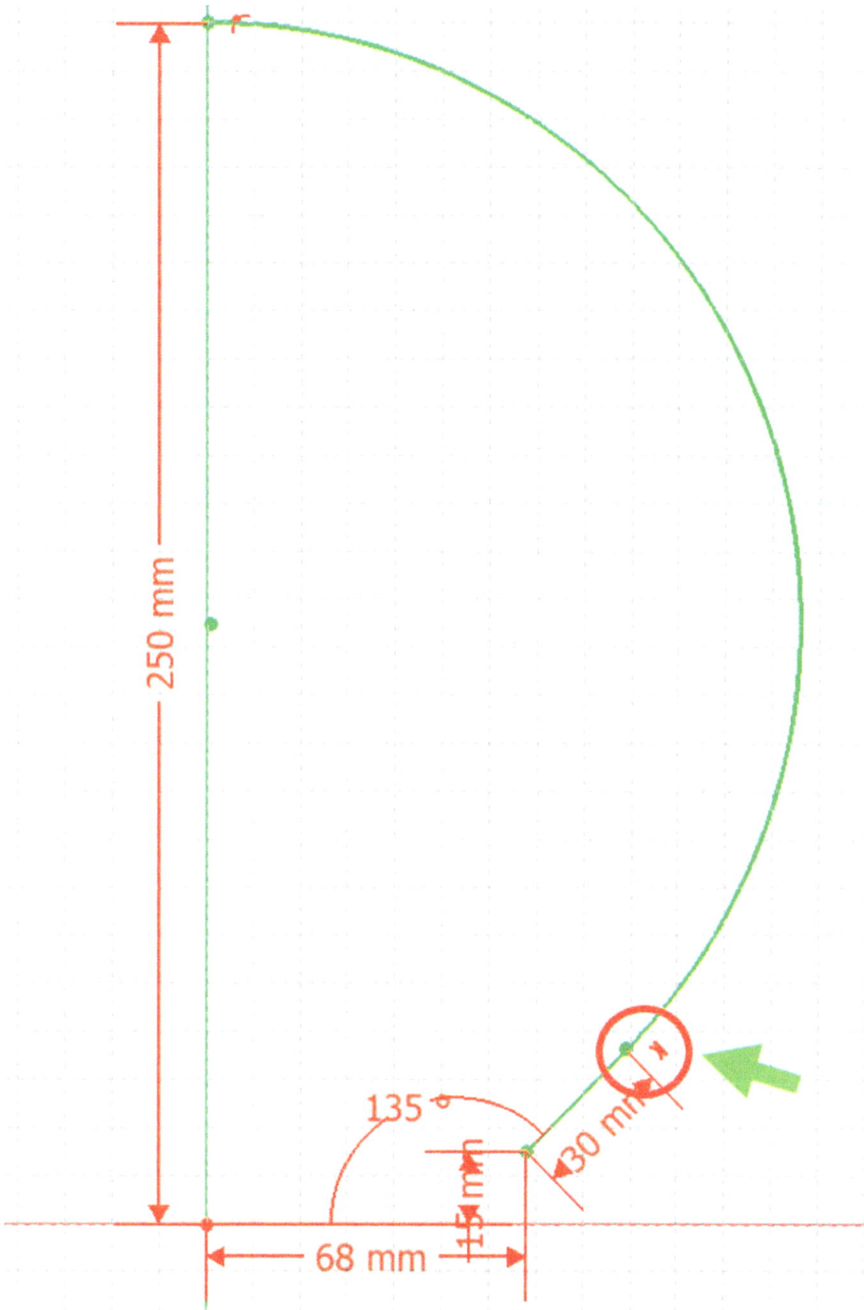

Then we add another line in the front area of the watering can. The line should be 40 mm long, have an angle of 30° to the horizontal sketch axis and a vertical distance of 140 mm and a horizontal distance of 66 mm to the coordinate origin.

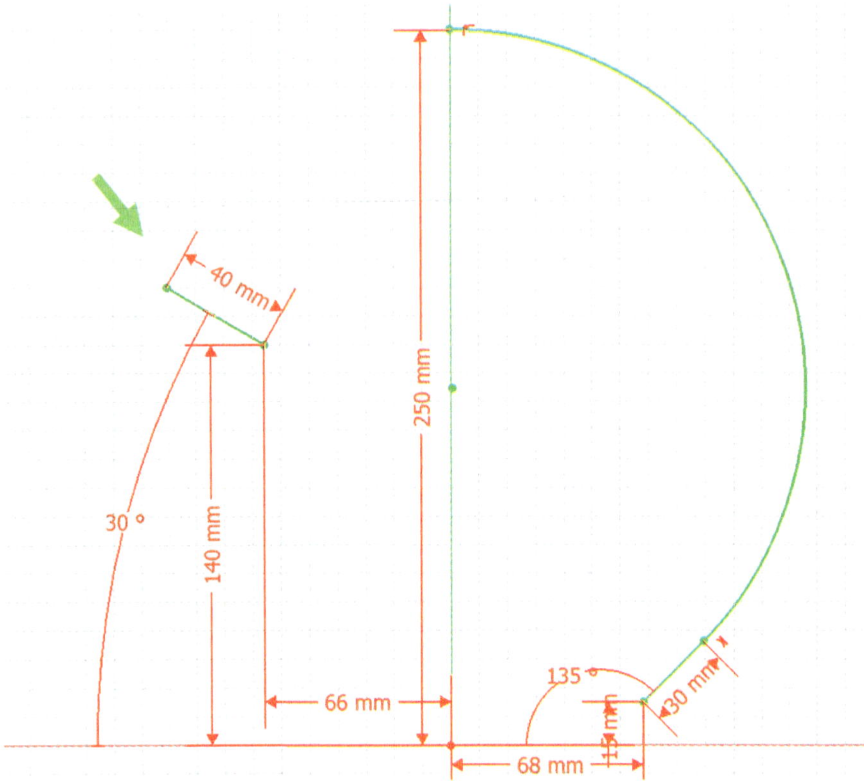

We will complete the path with another 3-point arc ("End points and rim point"), which we will set in the area of the vertical sketch axis tangential to the first 3-point arc.

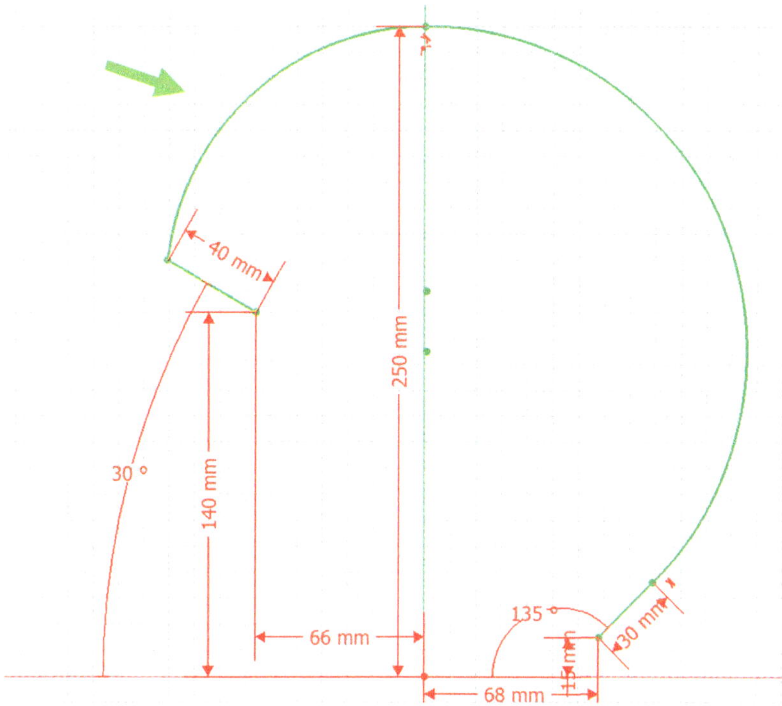

We round the front area with the command "Constraint-preserving sketch fillet".

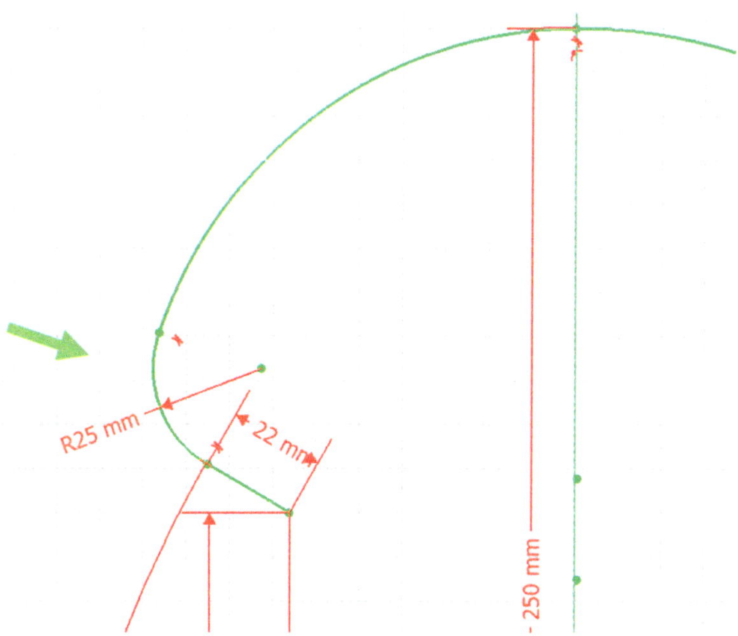

For the fillet, we dimension the resulting arc with a radius of 25 mm and the length of the shortened line segment with 22 mm. Then the sketch is completely defined and can be closed.

After we have finished the sketch, we can create the handle using the "Additive Pipe" command. To do this, we first select the profile we drew for the shape of the handle (ellipse) and then click on the command in the toolbar. Then we click on the button "Object" in the settings to select the path in the drawing layer.

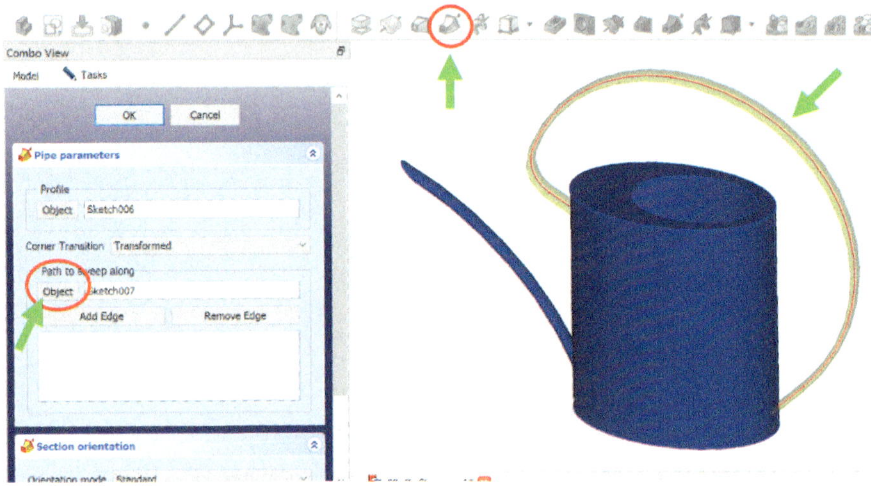

By the way: If an error comes, you can click it away. The error comes only because the program does not recognize a profile as a path. But we select the path manually.

In the penultimate step, we create a few fillets as follows: 5 mm for the top and bottom edges, 2 mm for the edges of the sprue parts (just click on the shell surface of the base body) and 0.3 mm for the top edge of the pouring can spout.

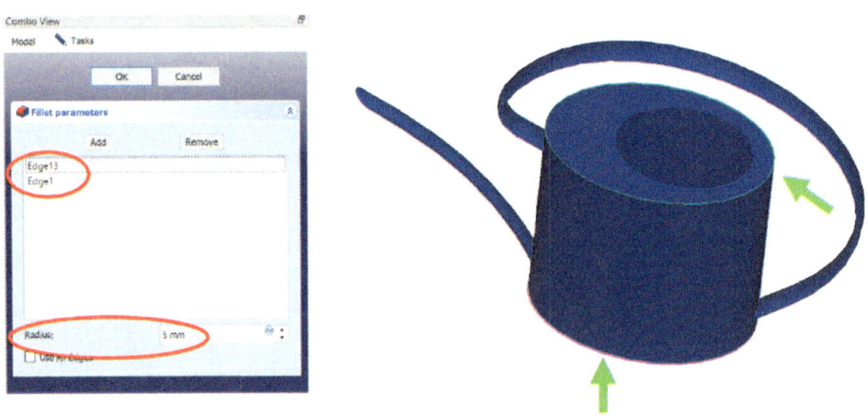

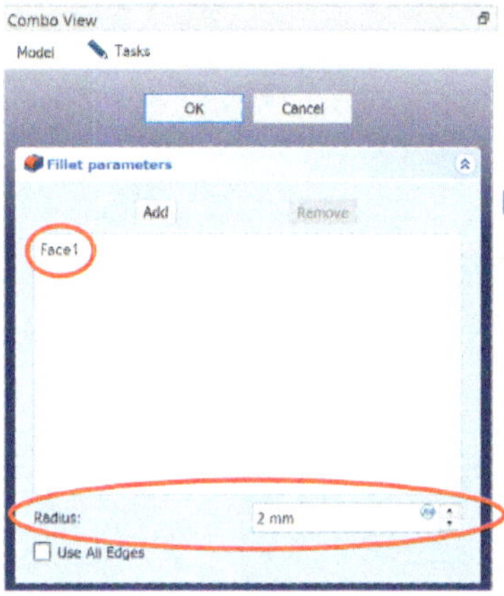

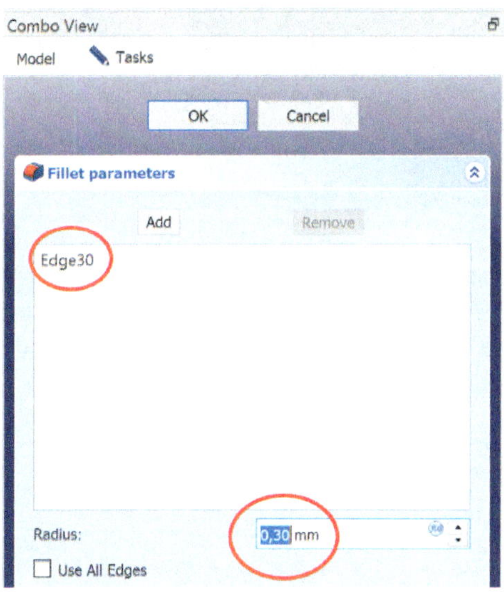

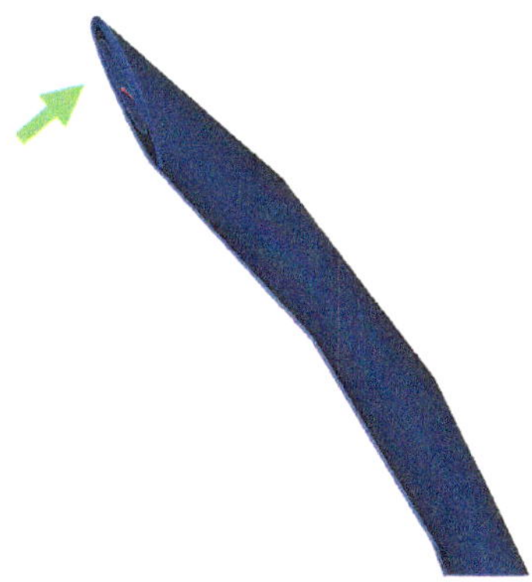

Excellent! As a final step, we would like to change the appearance a bit. For example, we could create a green plastic watering can.

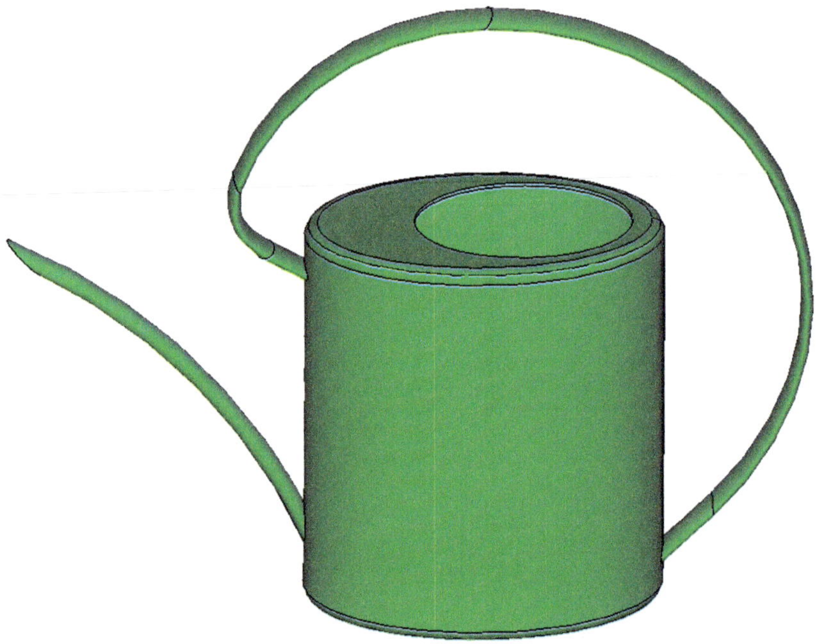

Now we have done it. The watering can is ready. By the way, you could print the 3D models we are designing here with a 3D printer. The best way to do this is to use my 3D printing beginners' course if you are interested in this topic! You will find information about this on the last pages of this book.

In the next chapter, we will design a housing of a remote control. Let's move on!

10 Project No. 9: Remote control

In this chapter, we want to design a remote control that will have a battery compartment with a slide-on cover, as well as some buttons. Normally, such a remote control is not constructed in one piece, but from several injection molded parts. In this case, however, we will design only one mockup, which we will make from two pieces.

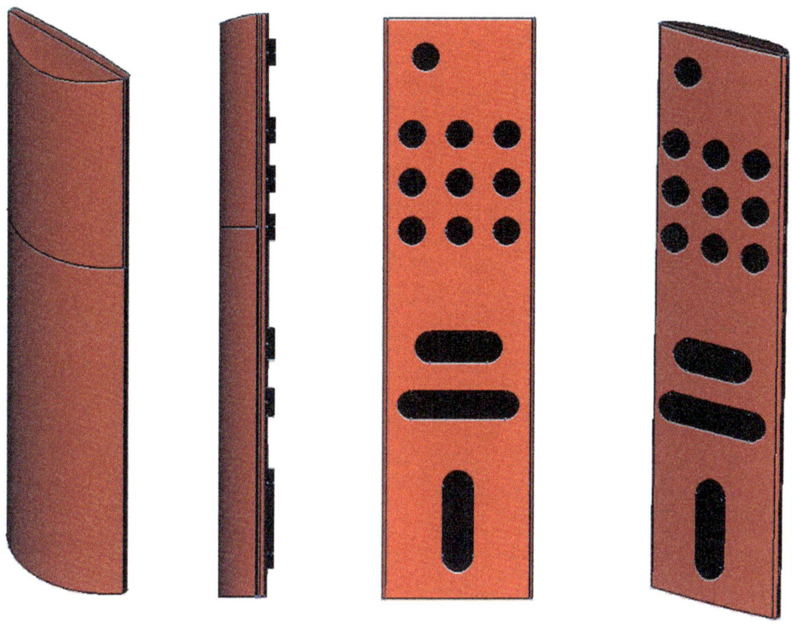

For the base body, which has an oval shape and which we will create using extrusion, we first need a 2D sketch on the x-y plane. We start for the shape of the cross-section with two 2 mm long vertical lines, one of which we place to the left and one to the right of the origin. The distance between these two lines should be 40 mm. The distance from one line to the origin should be 20 mm so that the lines sit symmetrically to the center line. Next comes a 3-point arc that connects the lower part and should have a radius of 25 mm.

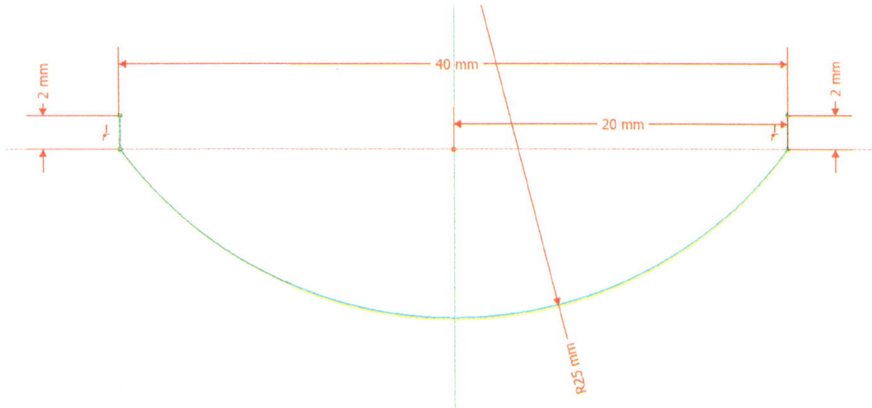

We place another sheet with a 200 mm radius on the upper side.

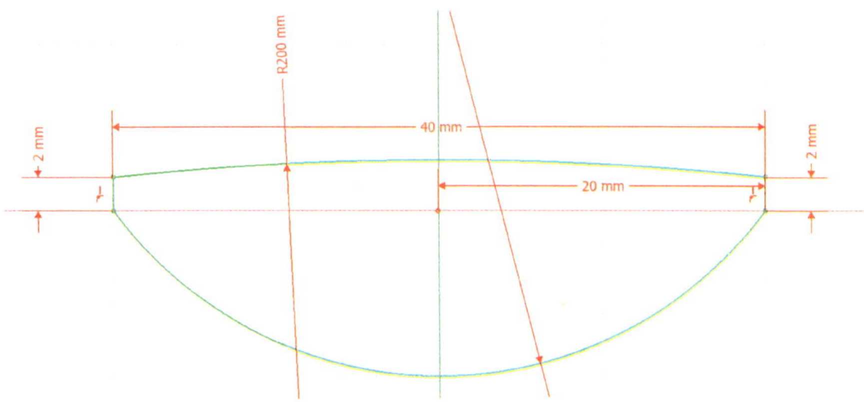

Then we can close the sketch and create the body by a symmetrical extrusion with 75 mm spacing each. For this, we use as usual the command "Pad" and the setting "Two dimensions".

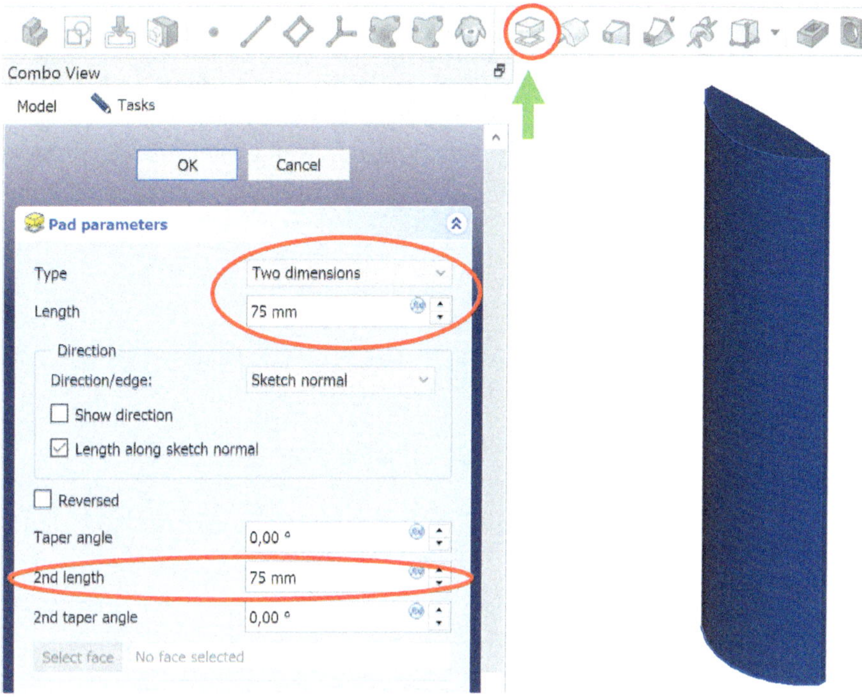

Then we round the four side edges with 1 mm each.

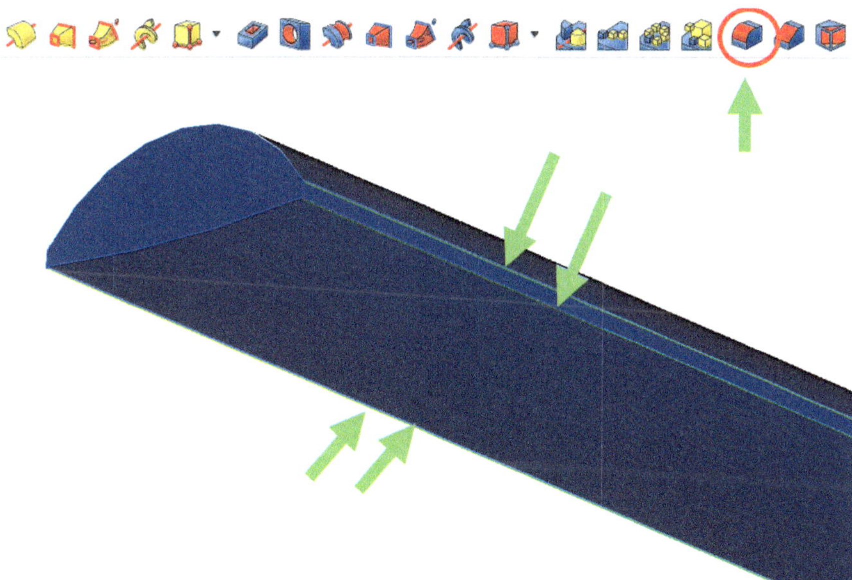

Then we want to bevel the upper surface of the remote control a bit. We do this with a profile on the side surface, which we then use to remove material from the base body. We draw the profile on the y-z plane in the upper area of the remote control. It is best to hide the base body for this purpose. The initial geometry is a horizontal line that starts on the vertical center line. This line should be 10 mm long and have a distance of 75 mm to the red horizontal sketching axis. At the endpoint of this line, we place a 2.5 mm long and vertical line.

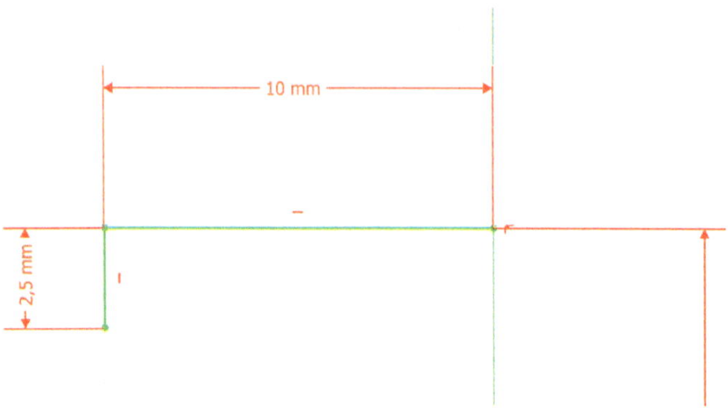

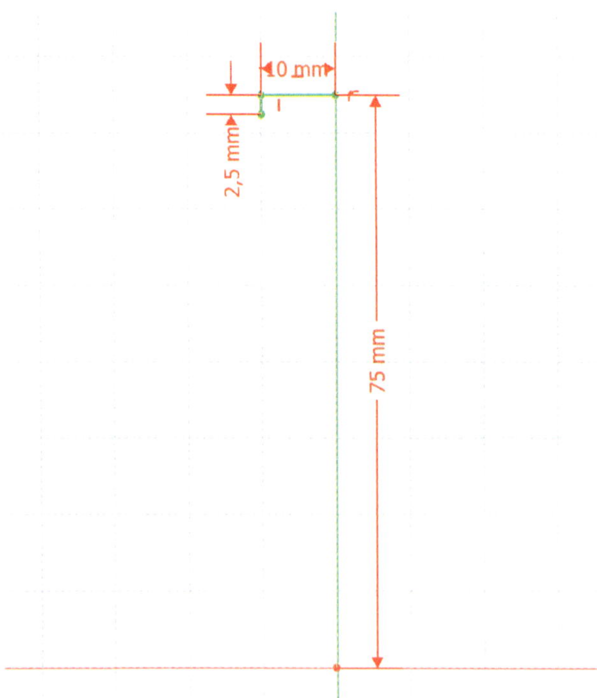

We link the two end points of the geometry with a 3-point arc. This is given a radius of 40 mm.

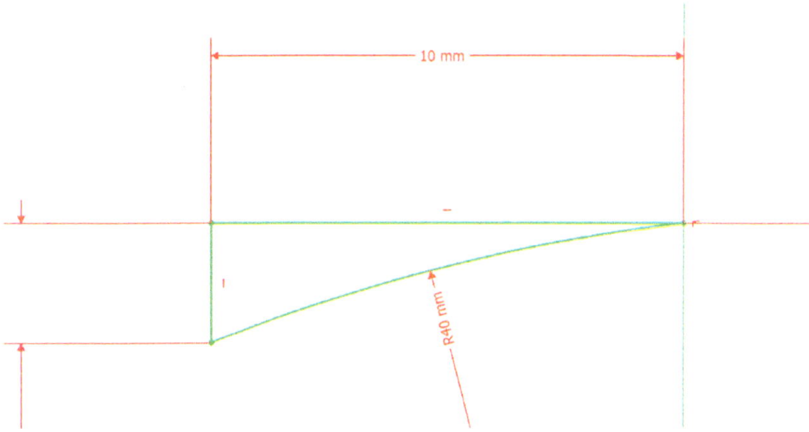

Now that we have sketched on the center plane, we need to cut out the profile from the center using the "Pocket" command, for example, 20 mm in two directions.

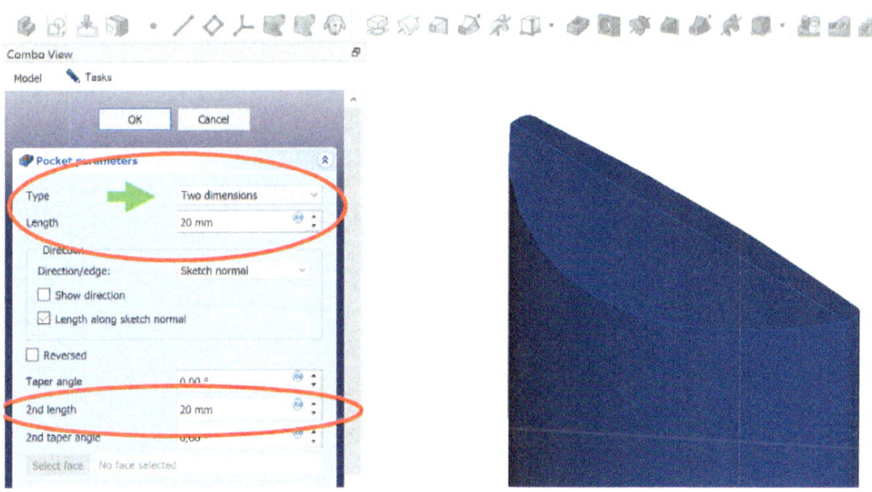

Next, we create a 20 mm offset plane to the x-y plane to create a sketch for the battery cover cutout on it.

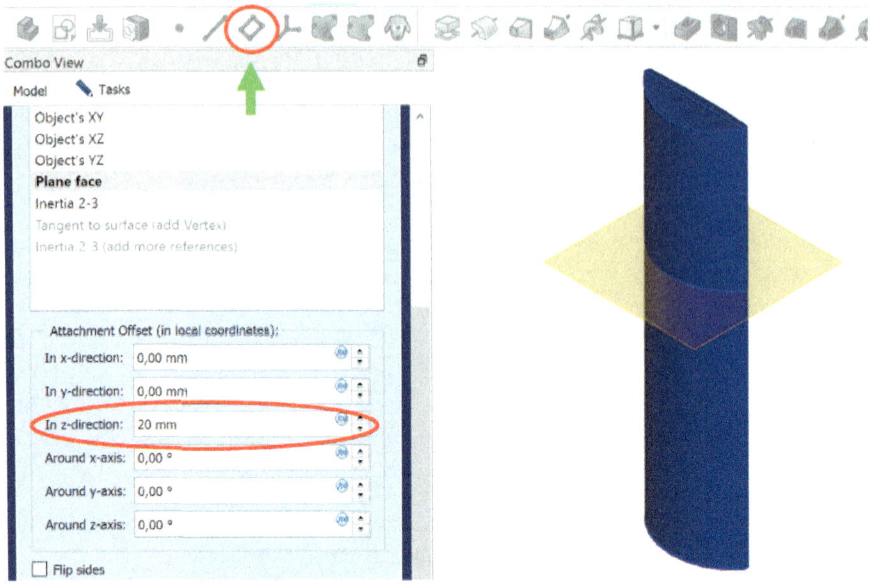

At this level, we outline the following profile:

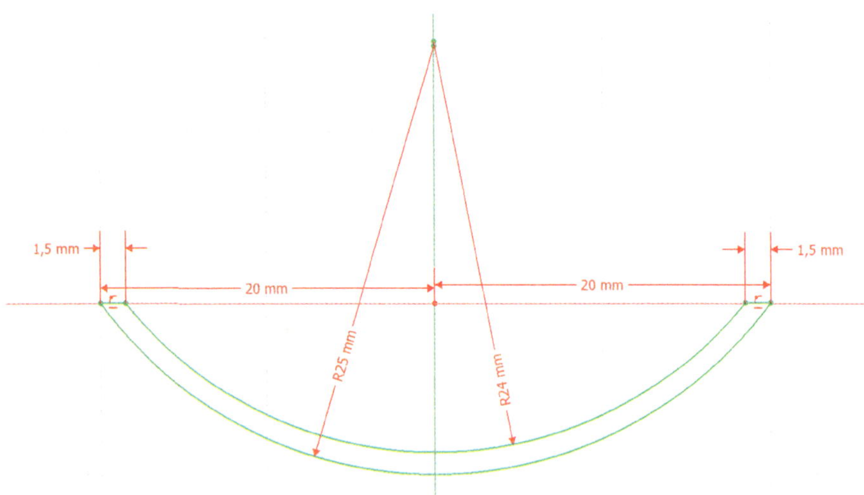

With this profile and with the help of the command "Pocket" we then remove material in the length of 95 mm (direction: downwards).

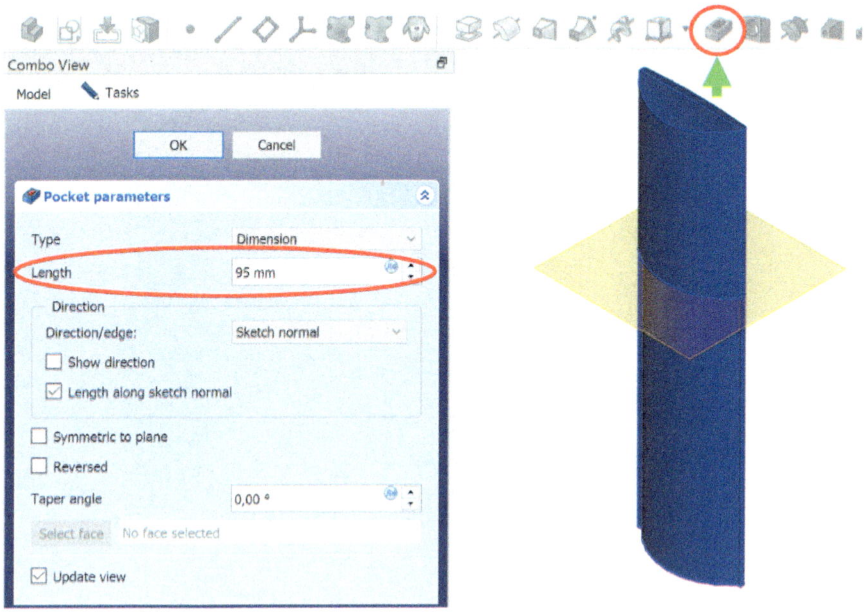

Next, we create a cutout to represent the battery compartment. To do this, we create a parallel plane to the x-z plane with a distance of 2.5 mm.

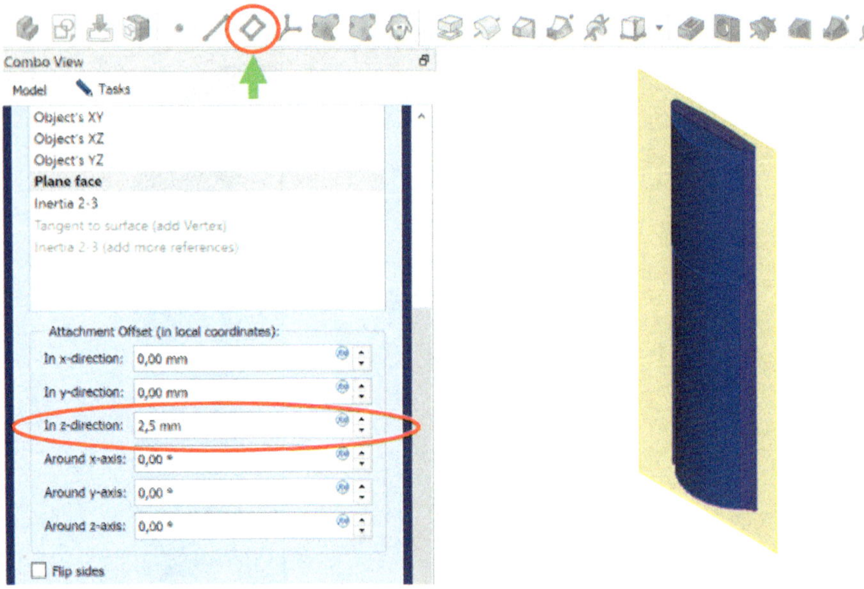

On this plane, we sketch the following rectangular profile:

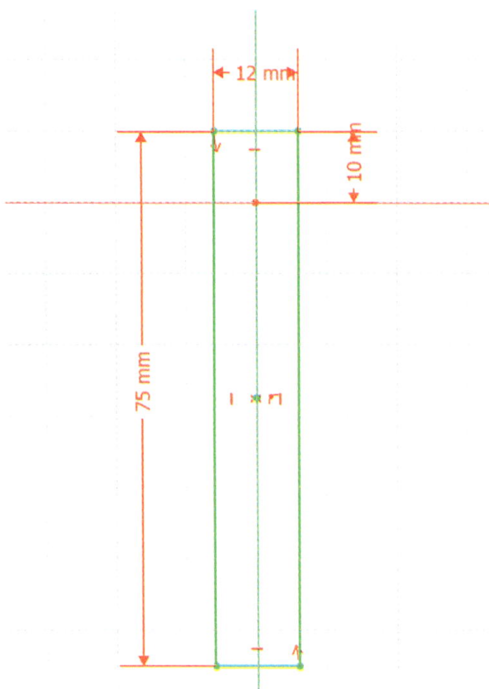

With the command "Pocket" we create the cutout (10 mm) and get the battery compartment this way. Probably you have to activate the option "Reversed" for this.

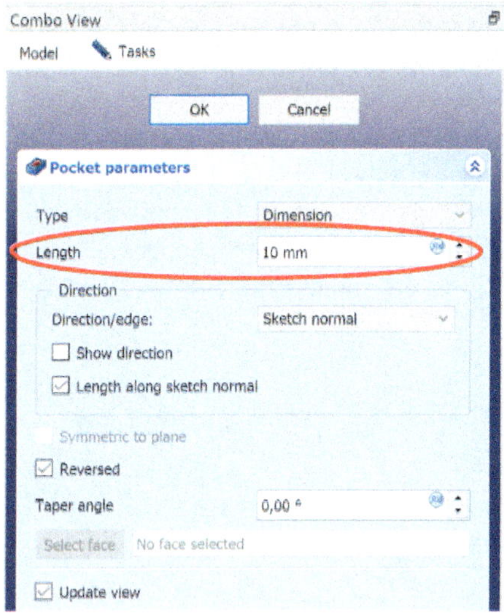

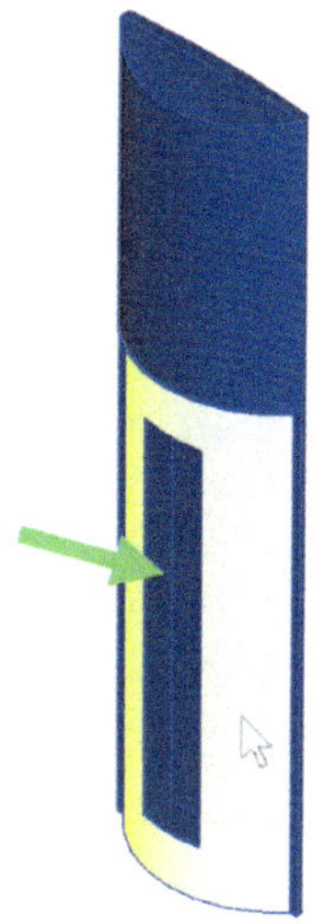

So that the front side doesn't remain as blank as it is at the moment, we now get to work on the sketches for the keys of the remote control. To do this, we create a sketch on the x-z plane of the solid. We extrude the buttons from inside the remote. We need to do this because the front face of the remote is curved. If we were to sketch on this curved surface, the side transitions of the buttons would not be linked to the surface. Just try this out for practice purposes, and you'll understand what I mean in a moment.

So, as before, we sketch on the x-z plane. For the first key, the on/off-key, we sketch a circle with a diameter of 7 mm in the upper-right corner and position the circle with 9 mm (horizontal) or 65 mm (vertical) to the origin. The next key also receives a 7 mm circle, which should be positioned with 12.5 mm (horizontal) or 45 mm (vertical) to the origin.

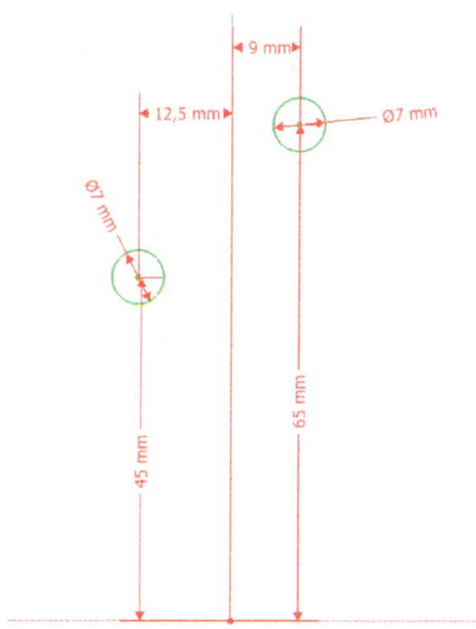

We will now create a keypad that will consist of a total of nine keys. To do this, we draw eight more circles, each with a diameter of 7 mm, as shown. To avoid having to dimension each circle, it is best to use the constraint "Constrain Equal" to make each circle identical to the first circle. There should be a distance of 12.5 mm between the centers of each circle. We also dimension these distances only once per row or column. To ensure that the sketch is completely defined, we also use constraints here. With the constraints "Constrain horizontally" or "Constrain vertically" we can set the circle centers horizontally or vertically to each other. We do this by first clicking on the command and then selecting the two centers of two adjacent circles.

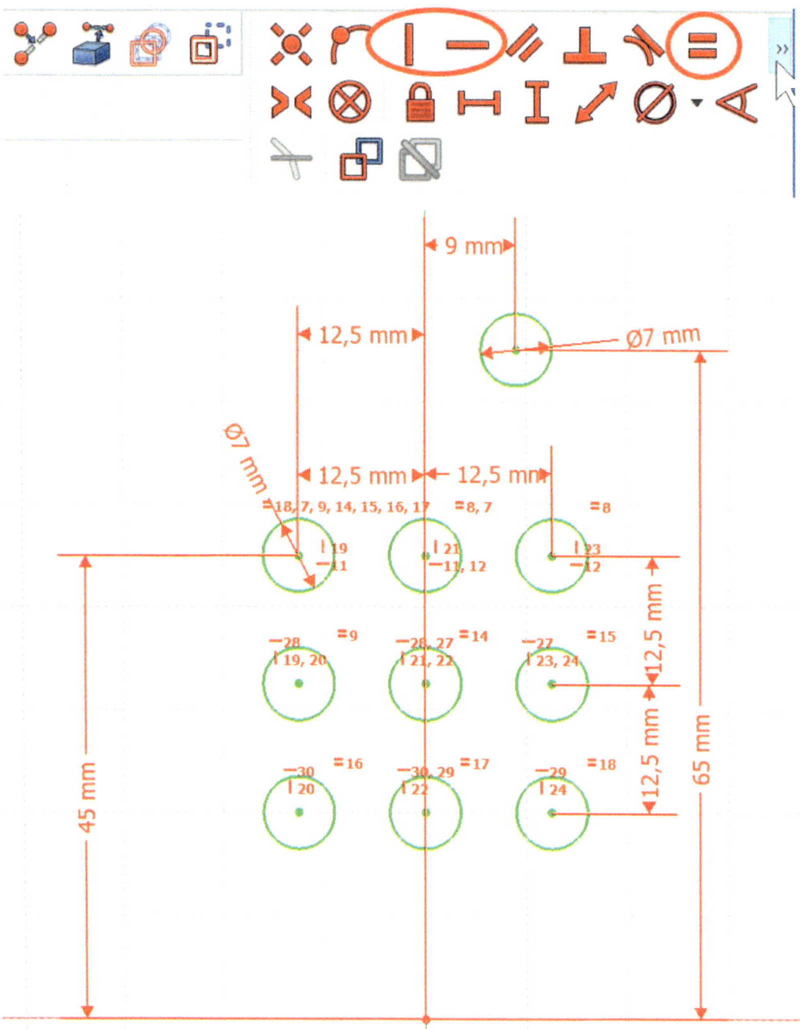

The last three keys are to be sketched as oblong holes. Two of them should be arranged horizontally, one of them vertically. This should look as follows, including dimensions. We are below the horizontal sketching axis, i.e., below the other keys.

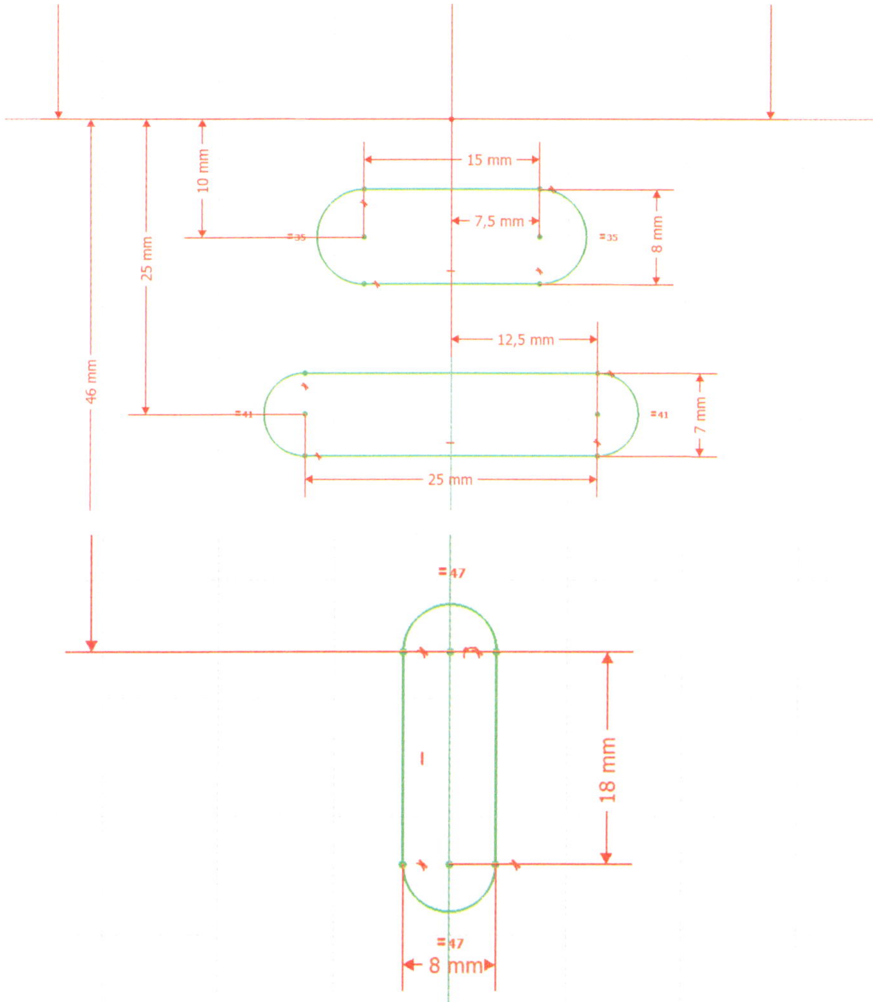

Then we can finish the sketch and extrude the keys 4.5 mm with the "Pad" command. Again, you may need to enable the "Reversed" option.

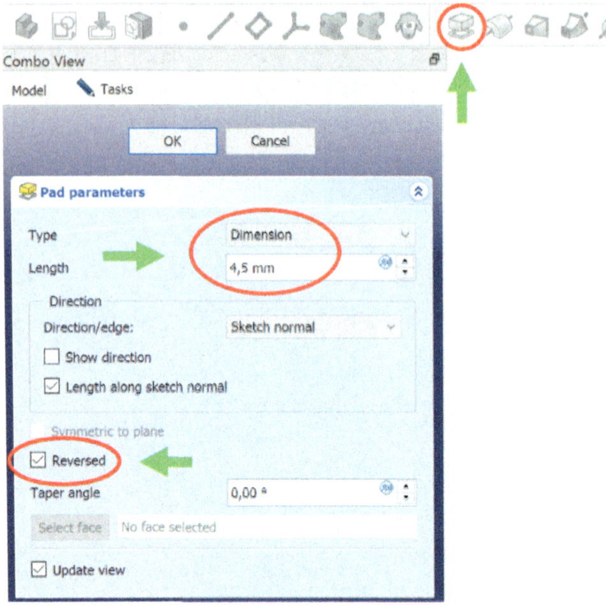

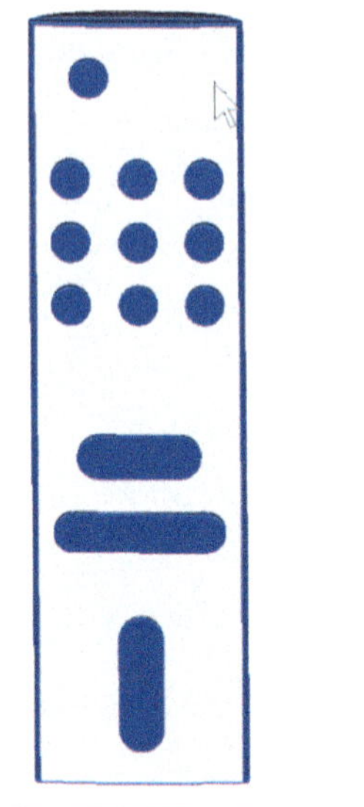

Since we are only drawing a mockup, the keys are connected to the case and are <u>not</u> functional, but we still want to color-code the appearance compared to the base body. We can color the base body in a red color, for example. We already know how this works from the previous projects.

On the other hand, we would like to color the keys black, for example. The procedure for this is a little different than usual. It is important that we first close the "Display properties" in the combination view in the tab "Tasks" with a click on "Close".

Then, we right-click on the feature of the keys ("Pad") and select the option "Set colors ...". With this command, we can colorize individual surfaces. We then select all the surfaces (buttons) that we want to colorize and then select a black color.

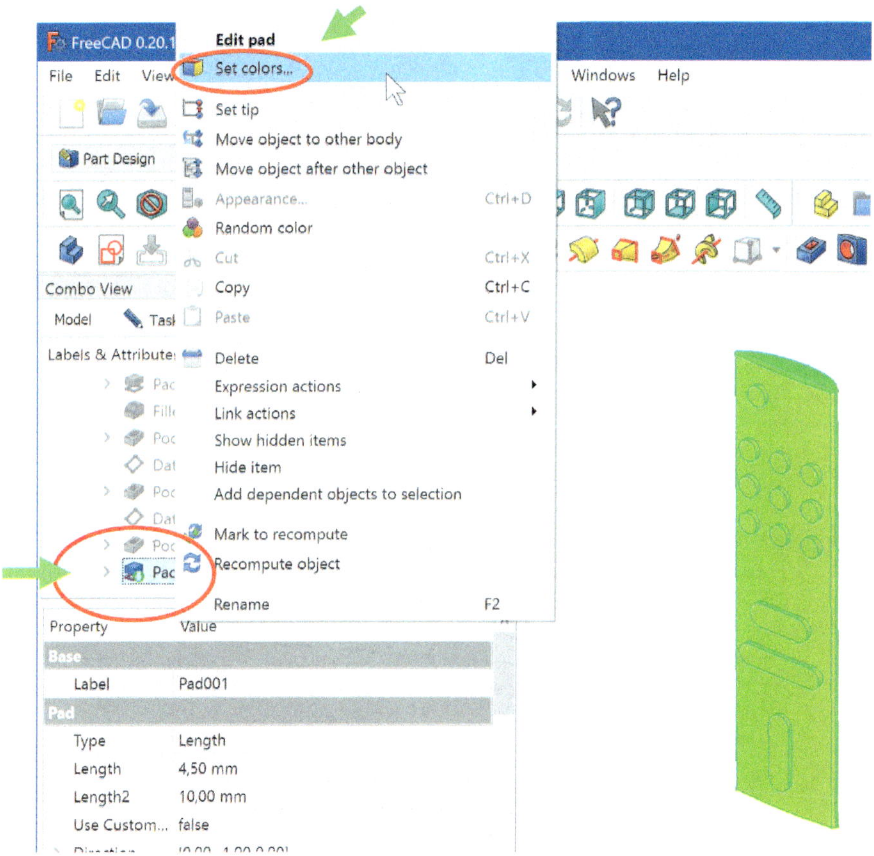

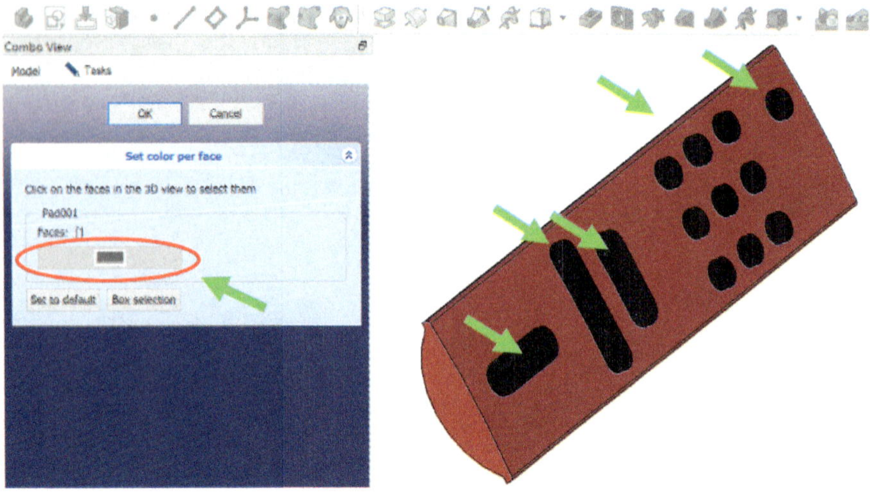

Now we are almost finished. In the next step, we would like to create an exact fitting cover for the battery compartment of the remote control. We will do that in a new document since this is a separate part.

In this new document, we create a body and a sketch on the x-y plane. On this sketch, we draw the profile that we already drew for the battery cover cutout.

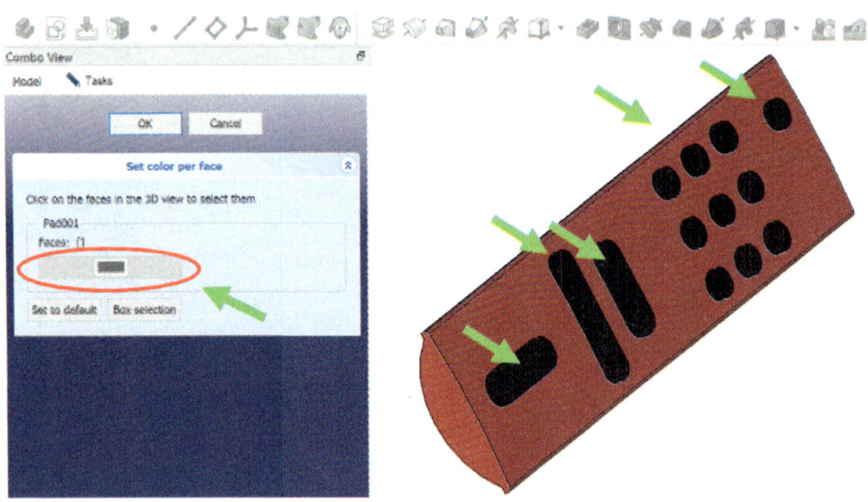

We then extrude this profile 95 mm and color it red as well. The part is now ready and can be saved.

Then we create a new assembly by creating a new document and switching to the "A2plus" workspace. In this workspace, we can mount the battery cover on the remote control. To do this, we add the remote control first and then the cover using the command "Add a part from an external file".

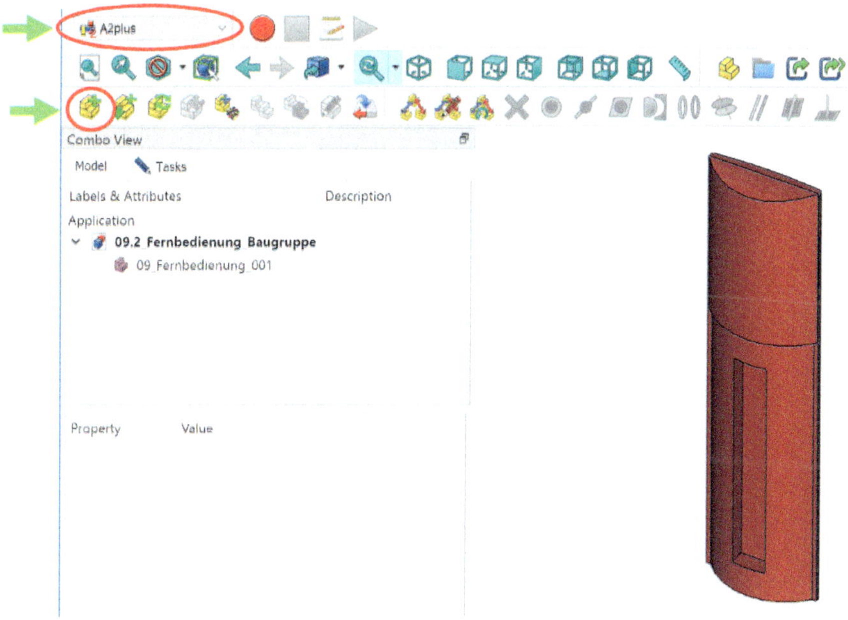

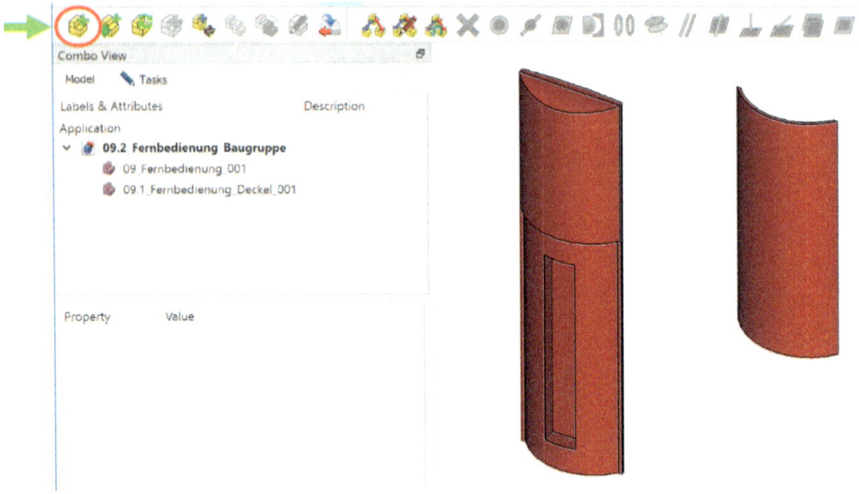

For mounting, we first select the inside of the lid and then the matching surface of the remote control. The possible constraints are then displayed. We select the command "Add axis Coincident constraint".

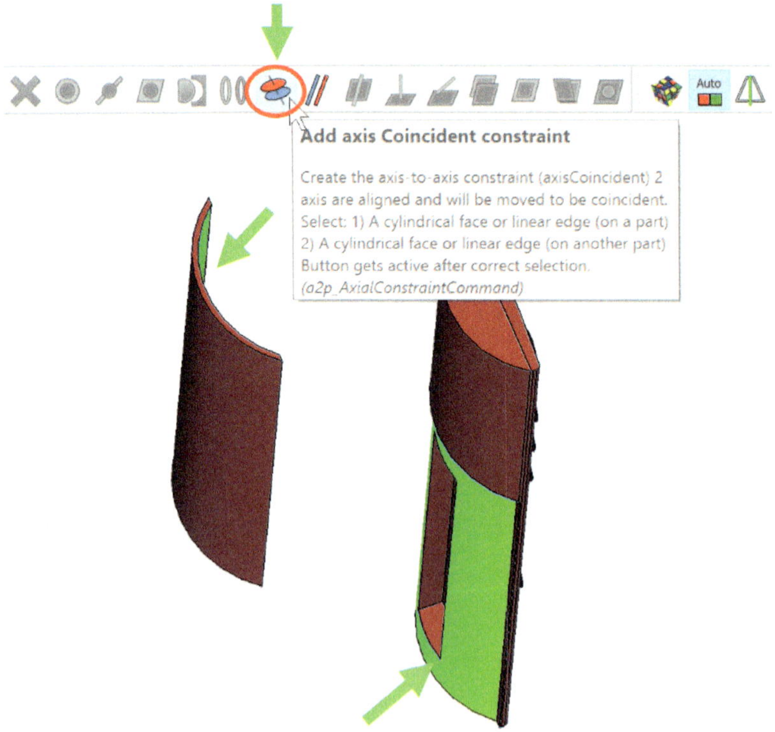

Now, the lid can still be moved in linear direction. In order to fix the lid in the correct position, we select the two displayed surfaces and click on the command "Add planeCoincident constraint".

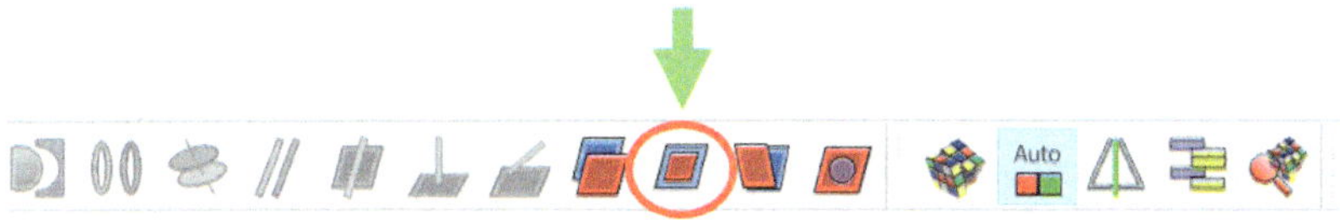

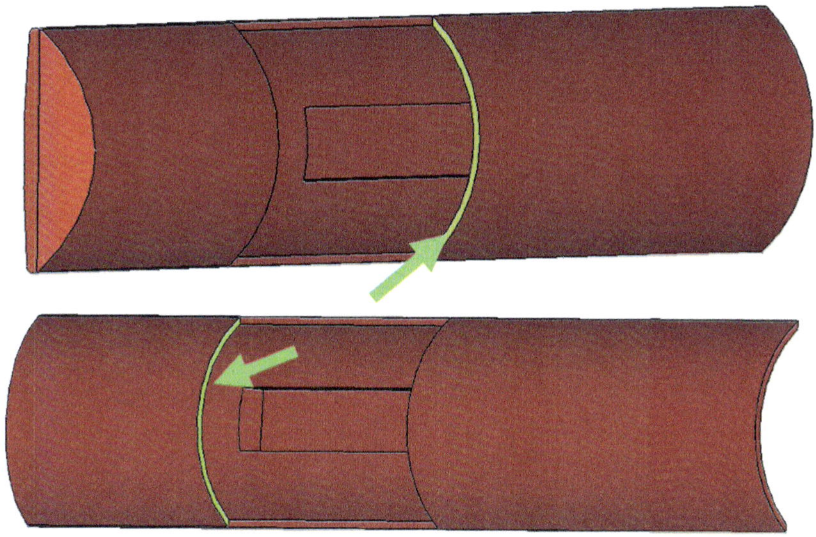

Perfect, now we can also successfully finish this design project and have reached the end of the course! Congratulations on making it this far!

Closing words

Excellent! You did it, with this chapter we finish the advanced CAD design course in "FreeCAD"! By now you should already have good skills in CAD design with "FreeCAD".

Together, we designed some great objects in this course, learned new functions and deepened basic functions. So, we've accomplished quite a bit! Be justifiably proud of yourself if you've made it to this lesson! Congratulations!

Eventually, there will be a second part to this advanced CAD design course, which will be similarly structured and will cover more moderately difficult as well as very complex design objects. If you are interested, please take a look at my author page from time to time to stay up to date.

And if you also want to experience your design objects for real in 3D, be sure to take a look at 3D printing. It's tremendously fun and useful to be able to materialize your designs.

The best way to do this is to use my course: "3D Printing | 101" and get started today!

If you liked the course on CAD design in "FreeCAD", I would personally be very pleased if you leave me a rating, as well as recommend the course! This will also help other interested individuals to make a decision. Thank you very much and see you soon.

Books on topics you might also like

All books are available online on the usual sales platforms. It's best to just search for the title, or feel free to visit my author page. Some of the books may not be published yet and will be released or found soon. Take a look at the books of your choice and your copy as e-book or paperback!

3D Printing:

CAD, FEM, CAM (3D Object Creation, Design, Simulation):

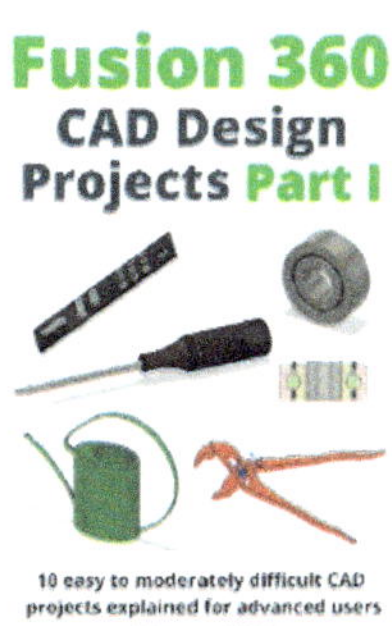

Electrical Engineering:

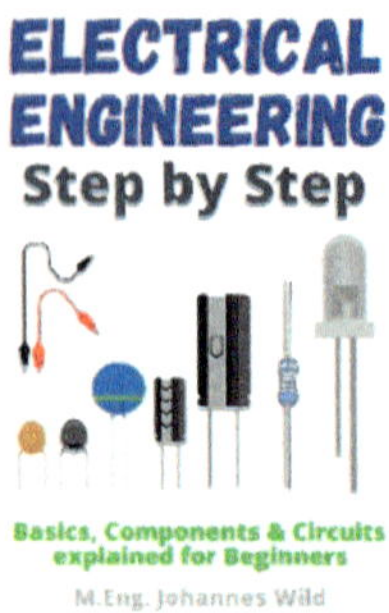

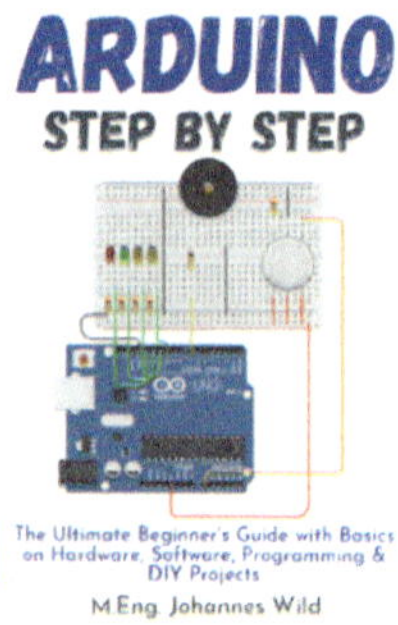

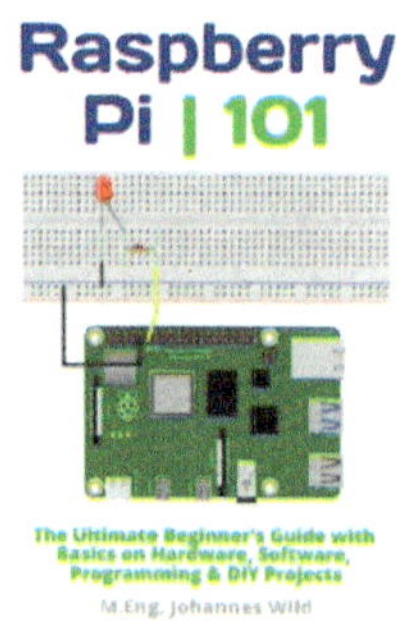

Programming and other Software:

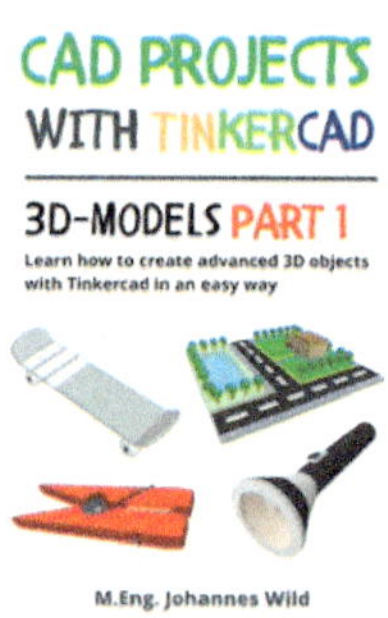

There are also identical video courses for some of these books:

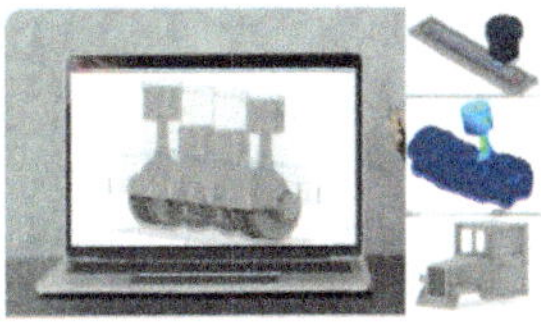

39 students 3h 33m

Fusion 360 Step by Step | CAD, FEM & CAM for Beginners

22 students 1h 18m

3D Printing 101 | The Ultimate Beginner's Guide - created by an engineer

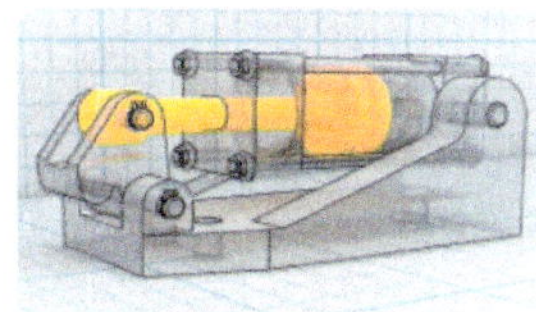

35 students 1h 18m

CAD Design 101 | 3D Modelling for Beginners - created by an Engineer

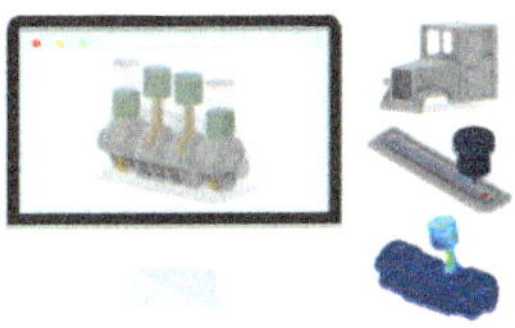

1 students 3h 33m

Fusion 360 Passo dopo Passo | CAD, FEM e CAM per principianti - da un ingegnere

1 students 1h 55m

Fusion 360 | CAD Design Projects – Part 1

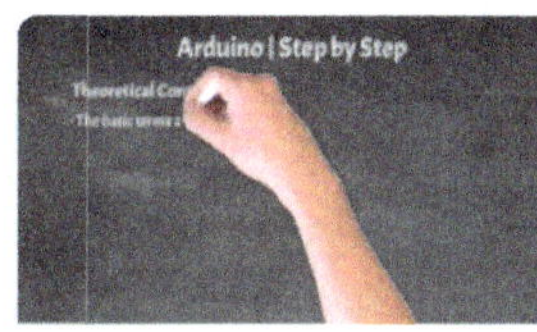

6 students 1h 53m

New The Arduino Course | Step by Step Explanations for Beginners by an Engineer

They are hosted on the learning website: skillshare.com

Be sure to use my following friends & family referral link to get a month of membership for free !

(I will get a little bonus if you choose to stay, so we will be both 😊 Thanks in advance!)

https://www.skillshare.com/r/profile/Johannes-Wild/854541251

It is best to copy the link in your browser to access the free month !

Sign up today and deepen your knowledge!

Imprint of the author / publisher

Johannes Wild
c/o RA Matutis
Berliner Straße 57
14467 Potsdam
Germany

Email: 3dtech@gmx.de

This work is protected by copyright

Thank you so much for choosing this book!

Printed in Great Britain
by Amazon

53147966R00076